OXFORD MEDICAL PUBLICATIONS

Liver Disease and Gallstones

THE FACTS

Liver Disease
and Gallstones

THE FACTS
Second edition

ALAN G. JOHNSON

Professor of Surgery and Honorary Consultant Surgeon,
University of Sheffield

and

DAVID R. TRIGER

Postgraduate Medical Dean, Professor, and
Honorary Consultant Physician,
University of Sheffield

Oxford   New York   Tokyo
OXFORD UNIVERSITY PRESS
1992

Oxford University Press, Walton Street, Oxford OX2 6DP

Oxford   New York   Toronto
Delhi   Bombay   Calcutta   Madras   Karachi
Petaling Jaya   Singapore   Hong Kong   Tokyo
Nairobi   Dar es Salaam   Cape Town
Melbourne   Auckland
and associated companies in
Berlin   Ibadan

Oxford is a trade mark of Oxford University Press

Published in the United States
by Oxford University Press, New York

First edition published 1987
Second edition published 1992

A catalogue record for this book is available from the British Library

Library of Congress Cataloging in Publication Data
Johnson, Alan G. (Alan Godfrey)
Liver disease and gallstones: the facts / Alan G. Johnson and
David R. Triger.—2nd ed. (Oxford medical publications)
Includes index.
1. Liver—Diseases.   2. Gallstones.   I. Triger, David R.
II. Title.   III. Series.
RC845.J64 1992   616.3'6—dc20   92-14362

ISBN 0-19-262305-2

Set by Advance Typesetting Ltd, Oxfordshire

Printed in Great Britain
by Biddles Ltd, Guildford and King's Lynn

Preface

The liver, to many, is a mysterious organ. Unlike the heart, brain, or kidneys, whose functions are fairly obvious, the liver seems complicated and its various activities confusing. The reason is that it has many different functions, from storage of carbohydrate to secretion of bile, from synthesizing blood-clotting factors to destroying drugs. This explains why, although machines have been devised to perform the main functions of other organs, (dialysis machines as artificial kidneys and cardio-pulmonary by-passes to function as heart and lungs), it has been impossible to devise an artificial liver.

In our own Western culture, the heart is considered to be the seat of the emotions—we talk about 'loving with all our heart'. In parts of Africa, however, the liver has pride of place, and in their language they will say 'my liver feels for you'!

Historically, the ancient Egyptians took special care of the liver when mummifying bodies. They placed it in a special jar with a human head on it. The Etruscans would inspect the livers of animals as omens and guidance from their gods, and in the Middle Ages, bile was one of the four 'humours' whose imbalance was thought to cause different diseases: the word 'melancholy' is still used of depression and is derived from two Greek words meaning 'black bile'.

Animal livers have been used for food for many years and are a rich source of carbohydrate, proteins, and vitamins. Before vitamins could be made artificially, the treatment of pernicious anaemia (a deficiency of vitamin B_{12}) was to eat many pounds of raw liver each day. No wonder many poor patients considered the treatment worse than the disease! On the other hand, the early explorers in the Arctic ate polar bear's liver, which is particularly rich in vitamin A, and died from vitamin A overdose.

The human liver is the site of many different diseases. The purpose of this book is to give the facts as they are known at present, in as straightforward and non-technical a way as possible. But any account of liver disease must include a number of technical terms which we have intentionally introduced and explained. We must, however, remember that our knowledge is advancing all the time and what appears to be true today may be shown to be only part of the truth tomorrow. We shall mainly refer to those facts that are universally agreed and accepted, and when details are mentioned that are only conjectures, this will be made clear in the text. In the first chapter, only the essential facts about the normal structure (anatomy) and function (physiology) of the liver will be outlined to enable the reader to understand the diseases and treatments which are to be discussed. Chapter 2 outlines the tests that are used in detecting abnormal structure and function. The later chapters will deal with broad groups of diseases and will answer the common questions that we are regularly asked by patients and relatives.

Sheffield
June 1992

A.G.J.
D.R.T.

Acknowledgements

We wish to thank Mr Pat Elliott and the Department of Medical Illustration at the Royal Hallamshire Hospital, Sheffield for their kind help in preparing the illustrations in this book. We are also grateful to Mrs Carole Stenton and Mrs Fiona Griffiths for typing this manuscript.

To Esther and Jennie,
for their encouragement and support.

Contents

1 Structure and function of the liver

The structure of the liver

The normal human adult liver weighs about 1¼ kg (2½–3 lb) and it is situated in the upper-right-hand side of the abdomen. Normally the liver is tucked neatly under the lower ribs so that only a centimeter or so projects below the rib cage (see Fig. 1). It has a relatively soft consistency and a smooth surface, virtually identical in appearance to the calf's or pig's liver that one sees in the butcher's shop. If, however, it is enlarged, the lower border tends to spread downwards and to the left in the direction of the umbilicus (navel). It is normally impossible to feel one's own liver. This is because the lower edge of the normal liver can only be felt when the abdomen is fully relaxed, and any attempt at self palpation inevitably results in tensing of the abdominal muscles. It is important to realize that other organs, such as loops of bowel, may be present and be felt in the upper part of the abdomen. The spleen is often enlarged when the liver is damaged, although many other conditions quite unrelated to the liver may also cause its enlargement. It too is beneath the ribs but is on the left-hand side and cannot be felt unless significantly enlarged (Fig. 1).

The liver is kept in position by strong liver (hepatic) veins that join it to the largest vein in the body, the inferior vena cava; this takes blood from the legs through the abdomen to the heart. The liver is also supported by a transparent sheet of tissue (the peritoneum), covering its surface and is attached to the diaphragm (which is immediately above it). This allows the liver to move each time a breath is taken. This mobility of the liver helps to protect it from physical blows to the abdomen.

The liver is made up of two separate lobes, the right and left lobe, each of which can function independently. A

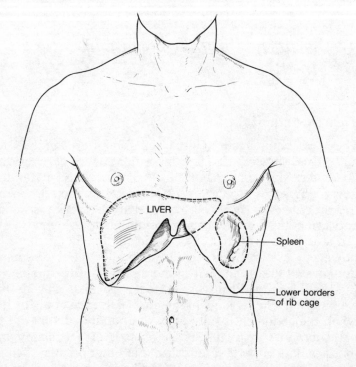

Fig. 1 The chest and abdomen showing the positions of the liver and spleen.

remarkable feature of this organ is its ability to re-grow if it is damaged. Occasionally victims of serious accidents may rupture their livers and may need to have as much as 75 per cent of the organ removed in order to stop the bleeding and to save their lives. Even after such major surgery the remaining liver is able to re-grow to its original size and shape within two or three months.

Another unique feature of the liver is that it obtains its blood supply from two completely separate and independent sources (see Fig. 2). Not only is it supplied by blood from the hepatic artery containing a lot of oxygen but also partly oxygenated blood reaches it from the stomach, pancreas, and intestines via the portal vein. This means that all the blood leaving the intestines reaches the heart and lungs only after

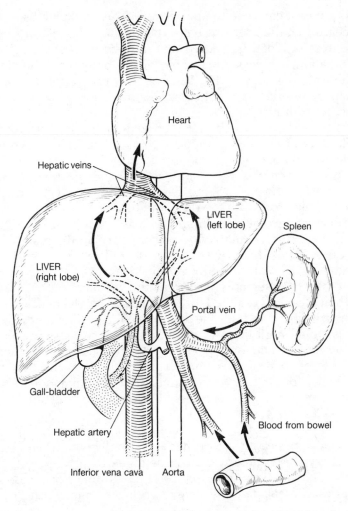

Fig. 2 The blood vessels of the liver. The liver obtains blood from both the hepatic artery and the portal vein.

passing through the liver, and this enables the organ to function as a very effective sieve in filtering food and other materials absorbed from the bowel. The total blood supply to the liver is about 1.2 litres (2¼ pints) per minute of which

80 per cent comes in the portal vein. This dual blood supply means that the liver is usually able to adapt if one of the two sources of blood is blocked or has to be tied off during an operation, whereas other organs in the body die if they lose their arterial supply. Blockage of both the portal vein and hepatic artery together however is invariably fatal within a matter of hours.

The blood leaves the liver by way of the hepatic veins. Blockage of these veins (hepatic vein thrombosis) is a serious but fortunately rare condition.

Apart from the veins and arteries the most important structures attached to the liver are the bile-ducts, which collect bile from both the right and left lobes of the liver and transport it to the duodenum, which is part of the intestine (see Fig. 3). The gall-bladder is connected to the main duct (common bile-duct) by the cystic-duct. The entrance to the duodenum is controlled by a sphincter or valve which allows the bile to enter only at certain times—mainly after meals when the bile is required for digestion. In between meals this valve is closed and the bile diverted to the gall-bladder where it is stored until the next meal. The sphincter also prevents food in the duodenum from regurgitating up the bile-ducts. The lower part of the common bile-duct passes through the pancreas on its way to the duodenum; inflammation of the pancreas may cause jaundice due to blockage of this duct (see Chapter 4).

The liver contains an intricate network of branches of the portal vein, hepatic artery, hepatic vein, bile-ducts, and cells that are closely interwoven so as to allow hundreds of different chemicals and molecules to pass from one to another in order to enable the liver to carry out its many functions.

The functions of the liver

The liver is an immensely complicated organ (second only to the brain in complexity) and many of its functions are only poorly understood. Here we shall concentrate on those which are most important in day-to-day function and also those that relate to liver disease.

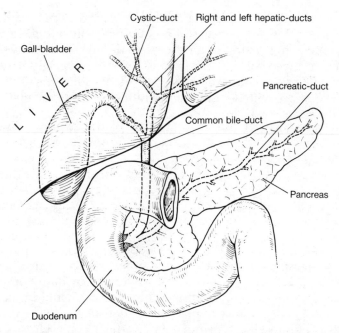

Fig. 3 The biliary system showing bile-ducts and the gall-bladder.

The conversion of food into energy

The most important function of the liver is its ability to break down the food we eat and turn it into energy; it is, so to speak, the major power-station in the body. The food we eat is made up of three major sources of energy: carbohydrates, fat, and protein. Each of these is partly broken down in the stomach and the intestine, from where it is transported directly to the liver via the portal vein.

Carbohydrates Sugar and starch (potatoes and bread), are the main sources of carbohydrate in the Western diet. These are broken down to glucose which is an important source of energy in the body. Glucose which is surplus to immediate energy requirements is stored as glycogen mainly in the liver and the body muscles. When required the glycogen can be rapidly mobilized and converted back into glucose.

Thus shortly after eating a meal we absorb a large amount of glucose, but the concentration of glucose in the blood is prevented from rising to an excessive degree by its conversion into glycogen and other substances. Conversely if we fast or starve for a prolonged period of time, the blood glucose is prevented from falling unduly by the conversion of glycogen back into glucose. This process is largely controlled by the liver with the help of insulin which is a hormone that is made in the pancreas. When the liver is badly damaged, however, this ability to control the glucose concentration in the blood is often lost. Patients with such liver damage often have to be given regular sugar to prevent their becoming unconscious from deficiency of glucose in the blood.

Fat Each gram of fat provides 9 calories of energy compared with only about 4 calories from a gram of carbohydrate or protein. Our major sources of fat in the diet include butter, cheese, cooking oil, and animal fat. This is mainly broken down in the liver either to provide further supplies of energy or to make specific types of fats of which cholesterol and triglyceride are the best known. High blood concentrations of these fats can lead to coronary artery disease. Obese people deposit excess fat all over their body particularly in their skin and muscles but also in the liver. Large amounts of fat are also found in the livers of people who drink excessively, and alcoholic fatty liver is discussed in Chapter 6.

Protein Meat and fish are the major sources of dietary protein in the Western diet, although vegetarians are able to obtain perfectly adequate supplies of protein from foods such as eggs, cheese, and nuts. Proteins are broken down into amino acids by the body and these comprise the 'building blocks' from which the cells and tissues in our body are made. There are 22 different amino acids, but 10 of these are called 'essential' since we rely entirely on our diet for these, while the remainder can be made from these 10 if not readily available. A large amount of the dismantling and rebuilding of proteins takes place within the liver.

Excretion of waste products

The body has two means of getting rid of waste products. Many substances are removed from the bloodstream in the kidney and excreted in the urine, but in some cases the chemical nature of the substance is such that it cannot be cleared by the kidney. Such substances are then removed from the blood by the liver and pass into the duodenum and then into the bowel by way of the bile ducts. Bile is the name of the liquid secreted by the liver in which such substances are dissolved and carried through the bile-ducts. The problems related to bile, gallstones, and bile pigments, are described in Chapter 3.

Absorption of fats and vitamins

In addition to getting rid of waste products, however, bile also carries a number of substances which assist the absorption of fat and fat-soluble vitamins from the intestine. When the flow of bile is reduced or blocked, as sometimes happens with liver or bile-duct disease, absorption of fat may be reduced and the patient will pass fat in their motions. This is recognized by the fact that the motions are very bulky, smell extremely offensive, float in the pan, and are difficult to flush away. In addition to this, vitamins A, D, and K will not be absorbed because they are only soluble in fat. Vitamin A deficiency causes night blindness while lack of vitamin D may result in bones becoming very brittle and breaking easily. Vitamin K is an essential factor in making blood clot, and when this vitamin is absent patients bleed and bruise easily. Deficiency of these vitamins can readily be corrected by injections, and such injections are often given to people with certain types of liver damage.

Hormones and enzymes

Part of the manufacture and breakdown of many hormones takes place in the liver. The hormones are far too numerous and complex to describe fully, but certain important groups deserve mention here. The production of many of the sex hormones in the body is regulated by the liver and so women with liver disease sometimes experience loss of menstrual periods while men notice loss of body hair and impotence. The liver is also important in making enzymes and proteins, which are responsible for most chemical reactions in the body, for example those involved in the repair of tissues which are damaged by injury. This means that patients with liver disease who sustain accidents of any sort or who undergo an operation tend to take very much longer to recover than do healthy people.

2 How do we investigate the liver?

People with heart disease usually get pain when they exert themselves, while those with lung troubles generally become breathless. In both of these the severity of the damage can often be gauged by the amount of exercise it takes to produce symptoms. There is no such test with the liver. Failure of the liver does not produce easily defined symptoms. Even jaundice, which is the best recognized sign of liver disease, is very variable.

For this and many other reasons we have to rely heavily on a variety of different tests in the investigation of liver function.

Blood tests

There is no single blood test which can determine how the liver works. There are two different types of test: those which measure a particular liver function and damage, and those that help to diagnose the cause of liver damage. An example of the first type is the test which measures the level of bilirubin (an orange pigment), in the blood and thereby the ability of the liver to remove it (see p. 33). Another test measures the concentration in the blood of enzymes called transaminases, which are released into the bloodstream from liver cells when they are damaged. The level of these enzymes in the blood reflects the amount of inflammation in the liver. The ability of the liver to make protein can also be gauged by measuring blood levels of certain proteins of which albumin is the most important. Another measurement concerns a further enzyme called gamma glutamyl transpeptidase. The importance of this enzyme is that it is released from the liver in response to alcohol abuse, and so the test for this enzyme is widely used in assessing patients with liver problems related to alcohol. Although these tests are very useful in following the progress of patients, they are of limited value in diagnosing exactly what is wrong.

Examples of the second type of test are those which identify hepatitis viruses (see Chapter 5). Unfortunately there are only a few diseases for which specific blood tests exist and other more complicated investigations often have to be carried out.

Ultrasound

This test has made an enormous difference to the diagnosis of liver and gall-bladder disease during the last ten years. The principle of this technique is that a hand-held probe which emits ultrasound waves is pressed on to the skin over the liver and gall-bladder. These waves are bounced off the organs and a radar-like image of the structures is produced (bats use such a method for navigation). All sorts of abnormalities in and around the liver can be detected in this way. In the liver itself tumours can be seen, as well as cysts and abscesses. Obstruction of the bile-ducts leading to their dilatation can also be seen, as can stones in the gall-bladder. Ultrasound can also show cancer in the head of the pancreas, which is an important cause of jaundice (see p. 35). This is a very simple and safe test which involves no needles or pain.

Radio-isotope liver scan

Another way in which the liver can be seen is by injecting into a vein in the arm a chemical tagged with a radio-isotope, which is taken up by the liver. The patient then lies under a special radio camera, which detects the small amount of radioactivity emitted by the chemical and 'photographs' the size and shape of the liver, as well as certain abnormalities within the liver, such as tumours. It too is very safe; the amount of radioactivity involved is extremely small. Although widely used in the past, it has now been largely superseded by ultrasound and the newer forms of scanning (see below).

CAT scan (whole body scan)

This has nothing to do with domestic pets, but is the abbreviation for computerized axial tomography. This technique can be used to scan almost any organ in the body, including the liver and gall-bladder. Like ultrasound, it is completely painless to the patient. It is more expensive and less widely available than ultrasound and usually gives no additional information compared with the simpler procedure. Occasionally, however, it is used in addition to ultrasound when the latter does not give a satisfactory picture or when more detail is required of a particular area.

Investigation of the gall-bladder and bile-duct

It is often difficult to decide whether the jaundice or the abnormalities shown up by these liver-function tests are caused by damage to the liver, or to the gall-bladder, or to the bile-ducts draining the liver. While the tests described above are often very useful, other investigations are frequently used in addition.

Oral cholecystogram

In this test, an iodine-containing chemical is taken by mouth. The chemical is secreted by the liver, concentrated in the gall-bladder, and then shows up on an X-ray. Any stones within the gall-bladder may be shown up (see p. 17), but the test will not work if the liver is not secreting bile normally. The oral cholecystogram has now been replaced by ultrasound and is generally only used when the newer technique is not available.

Bile-duct investigations

These are more complicated than performing ultrasound or cholecystogram and involve injecting the bile-duct with a substance that will show up the bile-duct on X-ray. The simplest way of doing this is to inject into a vein in the arm an iodine-containing compound, which is secreted by the liver into the bile-ducts. This is called an intravenous cholangiogram. It does not give a very good picture and, like an oral cholecystogram, cannot be used if the patient is jaundiced. It is a dangerous procedure in those few patients who are allergic to iodine. It is seldom used today. A more direct method is to inject the chemical directly into the bile-duct either from above (through the liver) or from below (through the duodenum). In the former a fine needle is passed into the liver under local anaesthetic, and the chemical injected while the patient is being X-rayed. This procedure is known as a percutaneous transhepatic cholangiogram (PTC) (Fig. 4). The alternative approach is to pass a fibre-optic instrument (after the back of the throat has been numbed with a spray and the patient has been sedated but not anaesthetized) through the stomach into the duodenum, to identify the opening where the bile-duct and pancreas drain, and to inject a similar opaque dye up one or both of these ducts while an X-ray picture is taken. This is known

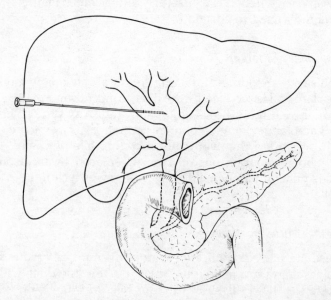

Fig. 4 Position of needle while X-raying bile-ducts via the liver (PTC).

as endoscopic retrograde cholangiopancreatography (ERCP) (Fig. 5). These tests are of great value in diagnosing the cause of obstructive jaundice as will be illustrated in Chapter 4. It is usually necessary to be admitted to hospital for these tests. Both of them are rather more invasive than the tests described earlier but even so they carry only a small risk of any complication.

Liver biopsy

This is the most important and direct investigation of the liver. Unfortunately it is more uncomfortable for the patients than most of the tests previously described. As described in Chapter 1, the liver lies just beneath the rib cage, and liver biopsy involves passing a needle between the lower ribs and sucking out a small piece of tissue. A local anaesthetic is injected into the skin overlying the liver so that the insertion

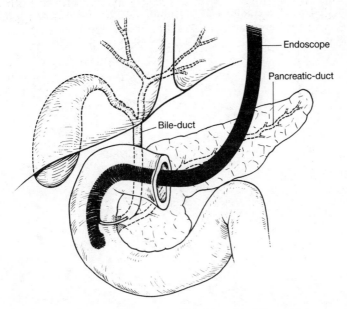

Fig. 5 Position of the endoscope while X-raying bile- and pancreatic-ducts from below (ERCP).

of the needle is relatively painless. The average liver biopsy specimen is less than 1 inch long (2 cm) and weighs only about 10 mg (one third of an ounce), but this is usually quite sufficient for diagnosis since most liver diseases affect every part of the liver.

Liver biopsy enables the doctor to tell how much inflammation and how much scarring has occurred. In addition, the pattern of liver damage often enables the doctor to decide what has caused the problem. The procedure has to be carried out with considerable care and is only done after simpler tests have shown evidence of liver disease.

It is usual to keep patients in bed for up to 24 hours afterwards since undue physical activity may cause bleeding from the liver. The patient is routinely checked beforehand for evidence of any tendency to bleed, since there is a risk of haemorrhage after liver biopsy. In addition to the chance of hitting a blood-vessel, one of the bile-ducts in the liver could also be inadvertently punctured. This rarely occurs unless the bile-duct is dilated, but when it happens the bile will leak into the abdomen and cause severe pain. For this reason liver biopsy is avoided whenever possible in patients with obstructive jaundice. Despite these complications the risk of serious problems after liver biopsy is less than one in a thousand.

3 Bile and gallstones

Bile

Bile ('gall' in older writings) is formed in the liver and secreted, via the bile-ducts, into the duodenum (see Chapter 1, p. 4). Its main constituents are bile pigments, cholesterol, and bile salts. It has a typically bitter taste ('as bitter as gall'), which can be distinguished from gastric acid if fluid is regurgitated or vomited.

Bile pigments

The yellow – green colour of bile comes from the bile pigments, which are called 'bilirubin' and 'biliverdin'. These pigments are formed by the breakdown of red blood cells and are derived from haemoglobin; in the liver they are joined (conjugated) to another chemical to make them soluble in water. The same change of colour is seen in a typical bruise when damaged blood cells release the red – blue colour of haemoglobin, which changes to green and yellow as the bruise resolves.

If there is excessive breakdown of red cells in the liver, then the pigment can solidify as 'pigment stones'. These are relatively rare and look like black coal-dust. They may be particularly troublesome because, being so small, they can easily be washed out of the gall-bladder into the main bile-duct. It is the bile pigments passing into the bowel that give the characteristic brown colour to faeces, so that if there is an obstruction to the flow of bile, not only does the patient become jaundiced (see Chapter 4) but the faeces become pale and lose their colour. In the bowel, the bilirubin is converted to another yellow pigment called 'urobilinogen', which is re-absorbed and excreted in the urine to give the characteristic colour of normal urine (Fig. 6). However, when the bile-duct becomes blocked, the bilirubin cannot escape by the usual

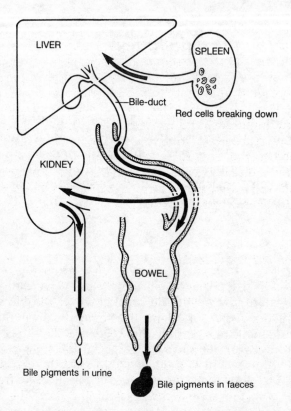

Fig. 6 The breakdown and destination of bile pigments.

route so it passes into the bloodstream from the liver and out in the urine, giving it a bright orange colour.

Bile salts

The second main group of constituents of the bile are cholesterol and bile salts (or acids). Cholesterol is a fatty substance which is commonly found in animal fats and is a normal constituent of bile. However, like other fats, it is insoluble in water, and so the only way it can be kept in solution is for it to be surrounded by molecules of bile salts

and phospholipids which are water-soluble. This arrange-
ment is called a micelle or 'wet crystal'. If there are not
enough bile salts to surround all the cholesterol molecules,
then they precipitate out and form cholesterol gallstones.
They are quite different from pigment stones and have a
waxy crystalline texture. They may be single or multiple (up
to many hundreds) and can grow very large. Bile salts are
normally reabsorbed from the lower small intestine and
returned to the liver to be used again. If this part of the
intestine is diseased or has been removed at operation, the
bile salts are lost into the faeces and therefore these patients
are more likely to develop gallstones from a deficiency of bile
salts.

Function of the gall-bladder

The gall-bladder acts as a reservoir for bile on the side of the
main bile-duct. It stores and concentrates the bile during
fasting periods and then contracts to expel the bile into the
duodenum in response to the arrival of food. Therefore the
maximum concentration of bile in the intestine coincides
with the time that it is most needed to aid the digestion of
fat. The signal for the gall-bladder to contract comes from the
release of a hormone ('cholecystokinin') into the bloodstream from the wall of the duodenum and upper small
intestine. If there is some problem with the chemical signal
or with the gall-bladder itself, it may not contract properly,
leading to stagnation of bile and an increased risk of
gallstones.

With this background, we are now in a position to answer
a number of questions about gallstones and their treatment.

Gallstones

The incidence of gallstones is increasing, and they are
the commonest reason for a major abdominal operation in
Britain—more common than appendicitis or duodenal ulcers;
in the USA it is estimated that 15 million people have
gallstones. Now that it is possible to detect gallstones by

simple ultrasound (p. 10), whole populations can be studied to find out the true prevalence of the condition. In Europe, 15 per cent of all females and 7 per cent of males between the ages of 18 and 65 years, have gallstones. The prevalence increases with age and at 65 years these figures have doubled to 30 per cent and 15 per cent. In other parts of the world such as rural Africa, gallstones are extremely rare, while in Japan there has been a great increase in cholesterol gallstones since the Second World War. There has been a well documented increase in incidence in Britain over the last twenty years. Whereas twenty-five years ago the typical patient with gallstones was described as a 'fat, fair, fertile, flatulent female of fifty' it is now not uncommon to see a small dark girl of 25 with the same problem. On the other hand, it is also increasingly common to find the complications of gallstones in very elderly patients with associated conditions such as heart and lung disease.

What are gallstones made of?

As we have seen they can either be made entirely of pigments or entirely of cholesterol, but in practice they often have a mixture with some pigment in the centre and cholesterol around the outside. It is possible that the pigment acts as the 'seed' around which the crystal grows, rather like a crystal hung in a supersaturated solution of copper sulphate starts crystal growing around it. In Britain, one in ten stones contain some calcium, which means they will show up on a plain X-ray (see below). In East Asia stones commonly consist of a combination of pigment and calcium (brown pigment stones), probably due to a different diet *and* parasites *or* infection in the bile-ducts.

What causes gallstones?

We have discussed the immediate causes above.

Black pigment stones These are common in two diseases in which the red blood cells are particularly fragile and therefore break down more quickly than normal. This rapid breakdown, known as 'haemolysis', usually causes anaemia, which is therefore called 'haemolytic anaemia'. The diseases are 'congenital spherocytosis' in white races and 'sickle cell anaemia' in those of West African origin. Interestingly, sickle cell disease actually protects the patient against malaria.

Cholesterol stones Although we know which chemical imbalance causes the cholesterol stones to form, we still do not know what causes this imbalance. The world distribution seems to follow life-style and possibly diet—being far more common in areas where the diet is high in cholesterol and low in fibre, but the story is probably not nearly as simple as that. Certainly people moving from an area with a lower incidence of gallstones to an area with higher incidence tend to acquire the incidence of the new area, suggesting the factors are in the environment rather than in themselves. For example, people emigrating from Britain and France to Canada have a six-fold increase in one generation, and those moving from East Asia to the USA soon have the same incidence as the rest of the population. But diet alone does not explain the variation within the high incidence areas and even those patients with excessive cholesterol in the bile seem to need some other particle or substance to trigger gallstone formation. Pregnancy seems to predispose to gallstones, either by altering the chemical composition of bile or by making the gall-bladder contract more sluggishly. Obese people are more likely to have gallstones than thin people, and rapid slimming is associated with a very high incidence of stones: if you rapidly lose stones, you rapidly gain stones! A short course of oral bile salts (see p. 24) over this period may help to prevent them.

Can gallstones be prevented?

At present the only advice that can be given is to cut down on the amount of animal fats and increase the amount of fibre in the diet. Gallstones disease is one of the few common conditions that does not seem to be related to smoking or alcohol intake!

Do gallstones run in families?

It is always difficult to establish that a disease is familial when it is very common in the general population. Having said that, there are two examples of gallstones 'running in families'. Some of the Red Indian tribes of North America, notably the Pima Indians, have a very high incidence in young people, about 80 per cent, producing cholesterol gallstones by the age of 35 years. The second group are those that make the *pigment* stones due to the haemolytic anaemias mentioned above. These anaemias are usually hereditary. Apart from these examples, gallstones occurring in mother and daughter can be related as much to a common diet and life-style as to common genes. Certainly, there is no reason why a person should develop gallstones just because her mother had them.

What symptoms do gallstones give?

Most gallstones give no symptoms at all! Population studies show that 80 per cent of people have 'silent' stones and were unaware of their presence. A proportion of these patients will develop symptoms with time but most will remain symptom-free for many years and many never develop any complaints. We have no idea why gallstones that have been silent for years suddenly cause trouble. Even when patients do develop symptoms and gallstones are found, it is often difficult to be sure that they are the cause. Like varicose veins, gallstones are blamed for every sort of ache and pain. It is important to be as certain as possible, because no-one wants to undergo a major operation only to find that the symptoms are the same as before.

Pain The most typical gall-bladder pain is felt in the upper right or centre part of the abdomen and often radiates round to the back, below or between the shoulder blades. It usually lasts for over half an hour, and is not relieved by opening the bowels. This pain is probably caused when the gall-bladder contracts after a meal in order to squeeze the bile into the intestine; the stones will also be squeezed into the outlet of the gall-bladder, and sometimes cause an obstruction. As the bile cannot get out, the pressure in the gall-bladder will rise and give pain (Fig. 7B). Gall-bladder contraction is especially stimulated by fatty foods and a low-fat diet can relieve much of the pain. Should a stone escape into the common bile-duct (Fig. 7C) it travels down the duct and may cause excruciating pain that may last for many hours and need strong pain-killers. Because the exit from the gall-bladder is usually narrow, it is the smaller stones that tend to escape. So it is probably better (not that you have a choice!) to have one large rather than many small stones.

Indigestion Rather vague symptoms, commonly associated with gallstones, are 'flatulence' (distension, belching) and 'dyspepsia' (burning discomfort and nausea). These are not directly caused by the gallstones but are related to the function of the stomach and pyloric sphincter. However, about half the patients with these symptoms will lose them completely after the gall-bladder has been removed and another quarter will have some improvement. So, only one quarter will still have symptoms as bad as before operation, and will have to have a special diet or drug treatment (see below).

Other complications If the stone becomes lodged in the lower end of the common bile-duct, it causes jaundice and, sometimes, acute pancreatitis, because the duct from the pancreas usually joins the common duct at its lower end and may, itself, also be blocked (Fig. 7D). Acute pancreatitis is caused by the digestive enzymes—normally excreted from the pancreas into the intestine—being activated within the gland itself and digesting and damaging the pancreas. This is very painful and often serious.

If the stone becomes impacted in the neck of the gall-bladder, two conditions may result. First, the gall-bladder can become acutely inflamed (cholecystitis), 7A, due to irritation by the chemicals in the bile, followed by bacterial infection. This acute cholecystitis produces pain, fever,

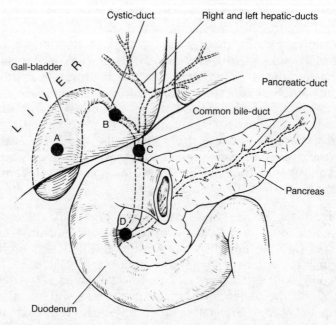

Fig. 7 The biliary system showing positions in which gallstones may give symptoms.

  A Stone in gall-bladder—causing mild or no symptoms.

  B Stone obstructing the cystic-duct—causing pain.

  C Stone obstructing the common bile-duct—causing pain and jaundice.

  D Stone at the lower end of the common bile-duct and pancreatic-duct—causing pain, jaundice, and pancreatitis.

nausea, and vomiting, and often requires emergency admission to hospital. Surprisingly, unlike the appendix, the inflamed gall-bladder hardly ever bursts. The second result of an impacted stone is less dramatic: the gall-bladder becomes chronically inflamed and may gradually distend with mucus or pus. This gives a general feeling of ill-health with poor appetite, often vague pain, and loss of weight.

How can gallstones be diagnosed?

As we have seen, a few gallstones contain calcium and will show up on a plain X-ray. The majority, however, require special types of X-ray to show them up (see Chapter 2). On an oral cholecystogram, which is rarely used, stones show up as shadows against a white background. Ultrasound shows the stones in the gall-bladder by the 'shadow' they cast. Techniques for diagnosing the cause of jaundice will be discussed in the next chapter.

Treatment

Can gallstones pass out spontaneously?

There is no doubt that small stones can pass out of the gall-bladder and down the common bile-duct into the duo-denum. It is unlikely that larger stones will pass this way and they are likely to cause jaundice or pancreatitis when they do.

Do gallstones need to be treated?

Apart from exceptional circumstances or when gallstones are found at another operation, no doctors recommend treatment for gallstones that are not giving any symptoms. A low-fat diet, which reduces stimulation of the gall-bladder, can often relieve mild or moderate symptoms and, particularly in an elderly patient, nothing further needs to be done.

Treatment of painful gallstones

Recently there have been several important advances in the treatment of gallstones. These can be divided into two groups: those that remove the stones and leave the gall-bladder, and those that remove both the gall-bladder and the stones ('cholecystectomy'). Although some of the new machines and techniques have been dubbed 'laser' methods by the media, lasers play little, if any, part in the treatment of gallstones.

Removing the stones but leaving the gall-bladder

The theoretical *disadvantage* of removing or dissolving stones, but leaving the gall-bladder, is that they could recur. In practice this is not such a major problem, particularly if the gall-bladder was functioning normally and not inflamed and the original stones were small. As pointed out above, some people may not be chronic 'gallstone formers' but have only developed them because of a specific life event such as pregnancy or weight loss. The theoretical *advantage* of these methods is that the gall-bladder is there for a purpose and if it is normal but just happens to have stones, it can be left to carry on its normal activity.

Dissolving with bile salts Small uncalcified stones in a functioning gall-bladder can be dissolved gradually by taking bile salts by mouth in capsule or tablet form. Two types of bile salt are used—chenodeoxycholic acid (a naturally occurring bile salt) and ursodeoxycholic acid which does not normally occur in man. Both are given at night and often in combination, the dose depending on the patient's weight. Both drugs have very few side-effects, although sometimes chenodeoxycholic acid causes diarrhoea and the dose has to be reduced. About 3 per cent of patients are suitable for this treatment and it takes up to a year's treatment to dissolve most stones. In practice these drugs are mainly used for patients with very small stones who have a contraindication for surgery, such as severe heart and lung disease, and stones may recur after treatment is stopped.

Shock-wave lithotripsy Over the last 10 years machines have been developed which can break up gallstones inside the gall-bladder by sending focused shock waves from outside the body. This technique has already been successfully used in treating kidney stones for many years. The aim of treating stones in the gall-bladder by lithotripsy is to break them into fine gravel which can easily be dissolved by bile salts. Although the treatment using the early machines was painful, newer models are painless and the patient is treated as an outpatient for a half-hour session and can go straight

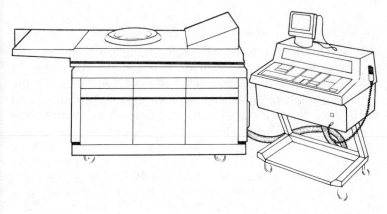

Fig. 8a Drawing of a Wolf lithotripter, showing the separate mobile control panel on the right.

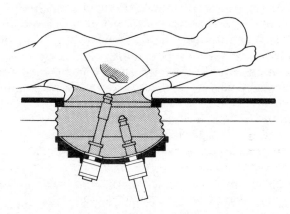

Fig. 8b Diagram of patient undergoing treatment, lying in contact with warm water. The gallstones are identified and the shock waves focused using ultrasound detection.

home. Usually, bile salts are given as well but some stones can be broken into such small fragments that the 'sand' is simply washed out of the gall-bladder. Symptoms are often relieved very quickly just by breaking up the stones—possibly because they no longer become jammed in the neck of the gall-bladder.

This is an effective treatment for patients with a functioning gall-bladder especially if the stones are solitary and less than 2.5 mm (1 inch) in diameter. In about 10 per cent of patients, stones recur over 3 years but it is easy to retreat the recurrent stone as no anaesthetic or incision is required. This method is an option to be considered particularly if there is any extra risk for operation, but it is suitable for 10–15 per cent of patients who wish to avoid both an anaesthetic and any scar.

Direct removal of stone ('cholecystostomy') The quickest way to remove gallstones is by inserting a small telescope or tube into the gall-bladder through a tiny incision in the abdomen (Fig. 9). There are various techniques, some using direct vision and some requiring X-rays, but the advantage over a major operation is that it can be done under a local anaesthetic. This is a great advantage for elderly, unfit patients or for those with other reasons for wanting to avoid an anaesthetic. The stones can be broken up and removed or dissolved. Sometimes a tube is initially left in for several days, but the patient does not need to stay in hospital. After an X-ray to check that no stones remain, the tube is pulled out.

Removal of the gall-bladder with its stones

Open cholecystectomy This has been the standard treatment for gallstones for over 100 years. It requires a general anaesthetic and an abdominal incision but the trend has been for smaller, more cosmetic incisions (Fig. 10), greatly improved postoperative pain relief, and much quicker mobilization and discharge from hospital, so that a young fit patient may only stay in for 2–3 days. Unfortunately too many patients are still told that they will have to be in hospital for 3 weeks and off work for 3 months! In experienced hands, it is a very safe operation indeed, with a mortality of less than 1 in 1000 for patients under the age of 50. Obviously in older patients the risks are a bit greater and the length of hospital stay longer, but anaesthetics and postoperative care have improved dramatically over the last 20

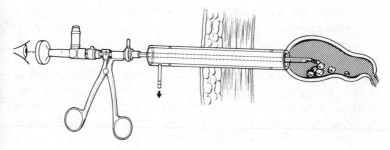

Fig. 9 Diagram of cholecystoscope, which fixes to the gall-bladder by suction, and through which gallstones can be removed under direct vision.

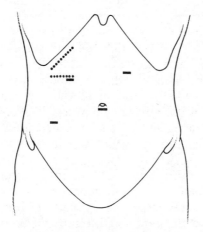

Fig. 10 Positions of commonly used incisions for open chole-cystectomy (dotted lines) and laparoscopic cholecystectomy (solid lines).

years. Patients are usually given an antibiotic at the time of operation to prevent any infection and an X-ray of the bile-ducts is often taken during the operation to make sure that no stones have come out of the gall-bladder unnoticed. If they are found, they can usually be removed at the same time.

Laparoscopic cholecystectomy ('key-hole' surgery) There has been great excitement about this new technique over the last two years but there is nothing mysterious or fundamentally new about it. Gynaecologists have used laparoscopic methods for operations such as sterilization for years. It is the development of very small TV cameras that has enabled very good views of the gall-bladder to be obtained. A telescope with a TV camera attached is introduced through an opening just below the navel after the abdominal cavity has been blown up with carbon dioxide gas to ensure a clear view above the intestines. Fine instruments are introduced through 3 or 4 other small incisions and the surgeon operates by 'remote control' (Fig. 11) while watching the TV monitor. The operation is more fiddly and sometimes takes longer than conventional cholecystectomy. Sometimes the surgeon encounters unexpected technical difficulties necessitating a change to the open operation described above, but this can be done without disturbing the patient. The main reason that many surgeons have changed to this method of removing the gall-bladder is that the small scars appear to cause less pain and lead to quicker recovery. Nevertheless it still requires a general anaesthetic and in a thin patient the 4 small scars may add up to the same length as one larger one (Fig. 10). The particular advantage may be in the fat patient, who would require a big scar for the standard procedure. The two techniques have not yet been thoroughly compared as far as operation time, safety, length of hospital stay, and patient comfort are concerned, but this is being done. While it appears that patients leave hospital very quickly after the laparoscopic method, the standard, open operation is still a perfectly acceptable and very safe procedure.

What happens if a gallstone is left behind at operation?

Occasionally a stone is left in the common bile-duct. This is rare nowadays because X-rays are used at operation and the inside of the duct is inspected with a small, sterile telescope known as a choledochoscope. Even if a small stone is left it does not necessarily mean a second operation. It may pass

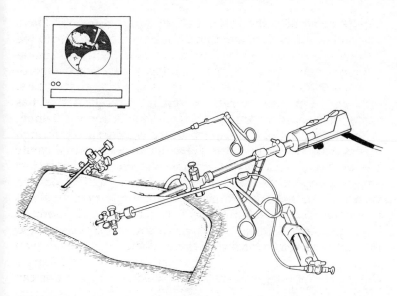

Fig. 11 Laparoscopic cholecystectomy, showing small TV camera attached to a 'telescope', and other operating instruments, through very small incisions in the abdominal wall. The surgeon sees the operation displayed on a TV monitor.

out on its own and there are now techniques to extract it through the opening made by the drainage tube or through the lower end of the duct via a gastroduodenoscope passed through the patient's mouth (see Chapter 4).

How long are patients in hospital and when can they return to work?

The simple answer is that they can go home and back to work as soon as they feel like it! Some patients react to the stress of anaesthetic and being in hospital more than others and the fatigue felt varies considerably. Even with open cholecystectomy, the muscles are sewn up with strong nylon which does not dissolve and so there is little risk of doing any harm even if people return to work very soon. The main

Liver disease and gallstones

activity to avoid in the first couple of weeks is heavy lifting, especially of awkward objects.

Is a special diet necessary after the operation?

Although it makes sense to avoid fatty foods before the operation, no such routine restrictions are required afterwards (although, of course, it is always wise to avoid a lot of fried food). About a quarter of patients will still have indigestion and have to restrict their diet but it is difficult to predict beforehand which these patients will be. It is best for most patients to aim to be back on a full diet in a few days and then just avoid anything that upsets them, trying it again, perhaps, some months later as the intolerance may be an immediate effect of the operation.

Does the loss of a gall-bladder matter?

The gall-bladder, as we have seen (p. 17), stores and concentrates the bile between meals so that concentrated bile is secreted into the intestine when fat is there to be digested. When the gall-bladder has been removed (or is blocked by a stone) a small amount of less concentrated bile is secreted into the intestine all the time, but there is still an increase with meals. This does not seem to matter in most patients, although some find they have to be careful about fatty food. Certain mammals, including the rat, do not have a gall-bladder—but most patients do not find this very reassuring!

Is there a link between gallstones and cancer?

Cancer can arise in the gall-bladder, nearly always in a gall-bladder containing stones and nearly always in patients aged over 65 years. Considering the huge number of patients with gallstones, it is a very rare tumour, but it can be very difficult to treat when it occurs. The two risk factors are very large gallstones and a lot of calcium in the wall of the gall-bladder, both of which can easily be detected by X-ray or ultrasound. These patients should have their gall-bladders removed even if their symptoms are slight, provided they are fit for operation and not very elderly. To remove all gall-bladders with stones just to prevent cancer would do more harm than good, and the incidence of cancer of the gall-bladder is actually decreasing in Western countries. There is no evidence that gallstones are associated with any other type of cancer, and equally there is nothing to suggest that removing the gall-bladder causes or prevents the development of cancer in other parts of the body.

Can gallstones be prevented?

As such a large proportion of the population has gallstones, it makes far more sense, both practically and economically, to concentrate research on preventing their formation rather than on finding more ways of removing or dissolving them. This is indeed being done and a number of factors have been discovered. Diet is obviously an important factor, but we are not yet at the stage where we can identify any single component in the diet which induces gallstone formation; similarly, we do not have a simple tablet to prevent gallstones being formed. It has been suggested that the daily aspirin that many men are taking to reduce the risk of heart attack may also reduce the risk of gallstones, but studies of this are still in progress.

4 Jaundice

Jaundice

Patients often say that they have had 'yellow jaundice'. In fact there is no other colour jaundice—as the word means 'yellow'! It is caused by the bile pigment being deposited in the skin.

It is interesting that jaundice is often noticed by a member of the family, or a friend, rather than by the patient. It is the yellowing of the whites of the eyes that distinguishes jaundice from the fading sunburn after a Spanish holiday and also enables recognition of jaundice in Oriental people. True jaundice can be mimicked by certain drugs, for example, mepacrine, which is used to treat malaria, and by the occasional patient who becomes addicted to carrot juice! Some patients notice the change in colour to orange of the urine, which often accompanies jaundice, before the change in colour of the eyes, but the modern trend for coloured bathroom suites makes this observation unreliable!

As mentioned in Chapter 3, bile pigments are produced by the breakdown of haemoglobin from the red blood cells which are processed in the liver and excreted down the bile-ducts. Since jaundice is due to the build-up of bilirubin in the blood, it can be produced in one of three ways:

(1) excessive production, when too many red cells are broken down (haemolytic jaundice);
(2) disease of the liver itself (hepatic jaundice) when it cannot process the bilirubin;
(3) obstruction to the bile-ducts preventing excretion into the intestine (obstructive jaundice or surgical jaundice).

The rare blood conditions that produce haemolytic anaemia have already been mentioned (p. 18) and, in addition, certain drugs produce acute haemolysis. In this first type of jaundice the urine does not change colour, because only the bilirubin that has been through the liver is in the right state to pass out into the urine. Parasitic infections such as malaria can also produce severe sudden haemolysis.

Gilbert's syndrome is an example of the second cause of jaundice. It is a very common type of mild chronic jaundice and is described in Chapter 11. Similarly jaundice at birth is generally due to the liver not being sufficiently developed to process bilirubin (Chapter 11).

Any disease of the liver may lead to impairment of excretion of bilirubin. The important liver conditions will include hepatitis (Chapter 5), those due to alcohol and other drugs (Chapter 6 and 7), and cirrhosis (Chapter 8). Here we will concentrate on obstructive, or 'surgical' jaundice.

Obstructive jaundice

Anything blocking the main duct outlet from the liver can produce obstructive jaundice. There may be stones or parasites within the duct, tumours in the wall, or tumours pressing from outside. Characteristically, in obstructive jaundice, the urine is very dark brown, or orange, and the bowel motions are pale. This is because the bilirubin, which gives the normal colour to the stools, cannot reach the intestine, and instead it goes back into the bloodstream and out through the kidneys into the urine. The two commonest causes of obstructive jaundice are gallstones in the common bile-duct, or a tumour of the pancreas encircling the duct at its lower end as it passes through (Figs 7 and 12). It is important to distinguish these before operation as the treatment may be quite different. An immediate consequence of any obstruction to a duct is that the duct above becomes distended (dilated) due to an increase in pressure, in the same way that a soft rubber tube attached to a tap swells if the end is blocked.

Let us now consider some common questions about jaundice.

How can it be diagnosed?

If there is doubt about the presence of jaundice the level of bilirubin in the blood can easily be measured, as can the amount of bilirubin in the urine. Blood tests also help to distinguish the different types of jaundice. The best way, however, to distinguish between diseases of the liver and obstructive jaundice is by ultrasound. If it shows that the main bile-ducts are dilated this is clear evidence of obstruction; and it may also be able to diagnose the gallstone in the duct or the tumour in the pancreas. Before the surgeon undertakes an operation for obstructive jaundice, he needs to know as much detail as possible about the position and nature of the obstruction. If the bile-ducts are dilated, further detailed information can be obtained by percutaneous transhepatic cholangiography (PTC) and ERCP, as described in Chapter 2.

Why does jaundice cause itching?

The itching that commonly accompanies jaundice is due to the bile salts (as opposed to the pigments) deposited in the skin. It can be helped by certain drugs, but they are not always effective in obstructive jaundice.

Do gallstones in the common bile-duct always cause jaundice?

Gallstones can be in the duct for some time—perhaps years— without causing jaundice. It is only when they block the lower end of the duct completely that jaundice follows. Sometimes they act as a ball valve and pop up and down in the duct causing intermittent jaundice.

When is jaundice painful?

Pain in jaundice is of two kinds. As we have seen in Chapter 3, while a gallstone is passing down the common bile-duct it can be very painful. The second type of pain is a dull ache in the right side of the abdomen, which is due to stretching of the capsule round the liver as the liver expands. This can occur in hepatitis (see Chapter 5) and in any condition that causes rapid liver enlargement.

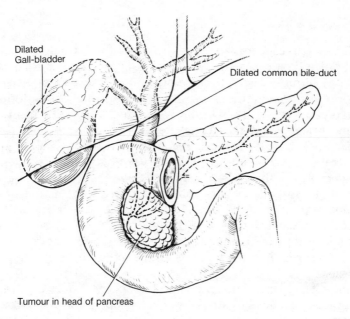

Fig. 12 Biliary system showing a tumour of the head of the pancreas obstructing the bile-duct.

How often is obstructive jaundice caused by cancer?

As we have seen in the previous chapter, gallstones in the common bile-duct are perhaps the commonest cause of obstructive jaundice, but cancer of the pancreas is also relatively common, particularly in older patients. The lower end of the bile-duct as it passes through the pancreas, is enveloped by the cancer (Fig. 12). The growth has usually reached a fair size by the time it blocks the duct completely. Less commonly a cancer may arise in the bile-duct itself, or in the duodenum just where the bile-duct enters. Occasionally, a secondary tumour may surround the bile-ducts as they emerge from the liver, or a primary cancer of the gall-bladder may encroach on the ducts.

Can obstructive jaundice due to cancer be cured?

Hardly any cancers can be removed by the time they cause jaundice. Just occasionally a small tumour in the duct or at the lowermost end can be removed and the patient cured. Some cancers of the pancreas can be removed, but this involves a very big operation with a high mortality and the cancer nearly always returns after a year or two. It is for this reason that most surgeons recommend a by-pass operation to relieve the jaundice without removing the tumour, especially in the elderly patient. Some cancers are slow-growing and the patient may live for two to three years. These operations are often quite simple and the patients recover quickly: they are blessed with such names as 'cholecystojejunostomy' or 'choledochoduodenostomy', but the exact replumbing depends on the individual tumour and patient.

Small tumours high up in the bile-ducts near or in the liver, can occasionally be cured by a large operation sometimes involving removal of part of the liver. Here again, a by-pass or hollow plastic tube through the tumour, to overcome the blockage, is often preferable.

What else, apart from gallstones and cancer,
can cause obstructive jaundice?

Not all tumours are malignant and sometimes benign tumours, which are curable, can block the bile-duct. Chronic inflammation of the ducts ('sclerosing cholangitis'), or of the pancreas ('chronic pancreatitis'), can also cause jaundice. It is sometimes very difficult to distinguish between a cancer of the pancreas and chronic pancreatitis, even at operation: this is another reason why a by-pass operation is preferred if there is doubt, as it cures the jaundice caused by benign inflammation and relieves the jaundice in cancer. A major, risky, operation on the other hand is unnecessary in the presence of inflammation and hardly ever cures cancer.

Cysts in the liver and pancreas can also press on the bile-ducts and cause jaundice.

Can obstructive jaundice be treated without an operation?

In Chapter 2 we talked about ERCP as a means of diagnosing causes of obstructive jaundice. This same procedure can also be used to treat the obstruction in one of several ways. The opening of the bile-duct into the duodenum can be widened by making a small cut internally which may allow a gallstone to be pulled out with a pair of forceps. This is a procedure which can be done through the endoscope with sedation (Fig. 5, p. 13). In patients where the obstruction is due to a tumour it is often possible to pass a small plastic tube with holes at each end down the endoscope, up the bile-duct and through the obstruction. This allows bile from the liver to enter the duodenum. This does not cure the cancer but does allow the yellowness of the skin to fade and also relieves any itching. This effect normally lasts for about six months, when the tube tends to get blocked; it may then be replaced unless the cancer has spread to other parts of the body.

How long does it take for the jaundice to fade once the obstruction has been removed or by-passed?

The jaundice does not fade immediately and may take three or four weeks before it completely goes. When it does fade, it leaves the patient looking particularly pale and 'washed out'.

5 *Hepatitis*

Introduction

Hepatitis means no more than 'inflammation of the liver'. It has many different causes: alcohol, viruses, and drugs being the most common. Hepatitis due to any cause produces very similar symptoms. Sometimes complicated tests are required to establish the cause of hepatitis, but in most cases the cause can be confidently established by appropriate enquiry into the patient's habits and activities. Alcoholic hepatitis is described in Chapter 6, while hepatitis associated with drugs and toxins is dealt with in Chapter 7. This chapter will consider the different types of virus hepatitis.

What causes virus hepatitis?

Virus hepatitis is very common throughout the world, affecting millions of people every year. It has been customary to name viruses after letters of the alphabet starting with hepatitis A, which was the earliest type to be recognized. Enormous advances during the last few years have enabled us to identify more and more viruses and each of these will be considered in turn.

Hepatitis A Hepatitis A or infectious hepatitis most commonly affects children and young adults, although people of any age are susceptible to it. It tends to be transmitted from one person to another by food, and it is particularly common where standards of sanitation and hygiene are poor, and where people are living together in a confined space. Hence, it is particularly common in schools, military camps, and overcrowded cities in underdeveloped countries. The last thirty years or so have seen a substantial improvement in the standard of living throughout Northern Europe and North America, and with this hepatitis A has become much less common. This means that epidemics of hepatitis are no longer occurring in our schools, but, instead, as people become older and start to travel to other parts of the world, they are at risk of getting hepatitis at a later age.

Hepatitis B Hepatitis B on the other hand is transmitted from one person to another by way of blood, hence its alternative popular name of 'serum' hepatitis. This means that the virus itself lives in the blood, and infection can occur if no more than a few drops of blood from an infected person enters the body of another individual. The most obvious way that this happens is through blood transfusions, but in developed countries all blood is screened for the hepatitis B virus before it is donated and this means of infection has been almost completely eliminated. There are however many other ways in which hepatitis B can be transmitted. Drug addicts who share syringes and needles for intravenous drug use commonly acquire hepatitis. Hepatitis B can also be sexually transmitted so it is very common among sections of the community who have many sexual partners, such as homosexuals and prostitutes. In Africa and East Asia, in particular, infection with hepatitis B commonly occurs at birth. The new-born baby may become infected with blood from the mother during labour. In certain parts of the world there are various rituals in which children or teenagers have cuts made on the body as part of initiation ceremonies, and the virus can be passed on in this way. Hepatitis B was first discovered in the blood of an Australian aborigine and it is for this reason that it is often called 'Australia antigen hepatitis'. It is particularly common among Australian aborigines because of these rituals. Another group of people at particular risk for developing hepatitis B are doctors and nurses, who may acquire it from their patients in hospital in the course of operating on them or dealing with them when they are bleeding.

Hepatitis C For many years we have suspected the existence of another virus which was responsible for transmitting hepatitis by blood transfusion which was neither hepatitis A nor B. For a long time we called it non-A non-B hepatitis, but during the last few years we have developed tests which are sufficiently reliable for us to describe this virus as hepatitis C. Although nobody has yet been able to see and grow this virus, the new techniques of molecular biology have enabled us to detect evidence of the virus in the blood. As with hepatitis B, we know that hepatitis C can be transmitted by blood transfusion and blood donations are now being routinely screened in most developed countries for this virus, although this process has only been introduced within the last year or two. Blood transfusion is not the only source of hepatitis C, as many cases appear to be quite unrelated to it. The way in which this virus can be transmitted from one person to another is not clear, however, and there is much debate as to whether or not it can be sexually transmitted. Hepatitis C is known to occur all over the world and is undoubtedly very important in certain countries such as Italy and Japan, although it is not thought to be a major problem in Britain.

Hepatitis D (delta hepatitis) The hepatitis D is a very unusual virus in that on its own it appears to be harmless but when it infects somebody who already has hepatitis B it can produce a very severe type of liver inflammation. Hence, if you are not infected with hepatitis B you are quite safe from hepatitis D virus. If, on the other hand, you already are infected with hepatitis B and you then pick up hepatitis D, or if you become infected with both viruses at the same time, then you risk having quite a severe infection which may progress to give cirrhosis (see later). Hepatitis D appears to be found mainly in parts of Latin America, the countries bordering the Mediterranean, and Eastern Europe, but with increasing foreign travel cases are now appearing all over the world.

Hepatitis E For many years we have recognized huge epidemics of hepatitis throughout Asia, Africa, and Latin America in which tens or even hundreds of thousands of people have been affected. Recently we have identified a new virus which we call hepatitis E, which has been responsible for most of these outbreaks. Hepatitis E is transmitted by contaminated water from wells or ponds and usually causes an acute illness which is not particularly severe except in pregnant women, when there may be a high mortality. To date there have been no epidemics in Western Europe or North America, but travellers may acquire it in the developing world and bring it back home with them.

In addition a number of other viruses can occasionally cause hepatitis, although they are more commonly associated with other types of illnesses. Glandular fever (sometimes called infectious mononucleosis) is the best known example of this. This is a very common infection in young adults, who usually develop a sore throat, fever, and enlarged glands in the neck. Another example is toxoplasmosis, an infection commonly acquired from dogs and cats, and which often resembles glandular fever. Occasionally, hepatitis can accompany a variety of other infections including measles and herpes simplex (cold sores).

Having excluded all these different viruses, we are still left with a few patients with hepatitis in whom we are unable to detect any cause. The diagnosis of a particular type of hepatitis depends upon specific tests for the virus; some tests are difficult to perform or there is a delay before the results become positive. Nevertheless, it is likely that more viruses will be discovered in the future and we will be moving further through the alphabet in years to come.

How does a patient get hepatitis?

Although there are many clearly identifiable sources of hepatitis as mentioned above, people often acquire hepatitis without knowingly being exposed to any of these recognized hazards. The most common example is contaminated food or drinking water in a café or restaurant. This risk is slight in richer countries with a high standard of hygiene, but is substantially greater in poorer countries with low standards of sanitation, where hepatitis is endemic. The problem of identifying the source of hepatitis is made greater by the fact that there is a long time period between acquiring the infection and developing the first symptoms. This time interval, known as the incubation period, is 5 weeks for hepatitis A and anything from 2 to 6 months for hepatitis B and several weeks for hepatitis C. With such a long incubation period it is understandable that most people have little idea how and where they are likely to have acquired their hepatitis.

What does it feel like to have hepatitis?

Hepatitis typically starts with an acute illness which is much like flu. The earliest symptoms are of general malaise, nausea, and loss of appetite, together with aching muscles and joints, high fever, and shivering attacks. This usually lasts for a few days and is followed by the onset of jaundice. At about the same time the urine becomes dark and the motions pale (see Chapter 4). Sometimes pain over the liver is experienced, but this is by no means a regular feature of hepatitis. The most uncomfortable period is usually before the jaundice develops; on average, the jaundice lasts for 1–2 weeks but sometimes can last a good deal longer. The major symptom that some people experience when they are jaundiced is generalized itching which is believed to be due to the irritant bile salts in the skin. For some this can be an extremely troublesome complaint, but several drugs are available to help this symptom.

The severity of the illness varies considerably from person to person. Most people probably acquire hepatitis without being aware of any illness, and certainly they are unaware that they have actually had hepatitis. In some, the jaundice which follows the flu-like symptoms is either very mild or is not detected, and in such cases the acute illness is likely to be labelled as flu. In general, children are much less severely affected than adults. On the other hand, a few patients may be severely debilitated for weeks, or even months, while very occasionally some die from acute hepatitis. The illnesses associated with the different hepatitis viruses may be identical, and they can only be distinguished by performing special blood tests. Hepatitis A, however, tends to be milder than B, and leads to full recovery, whereas a small proportion of patients with hepatitis B develop chronic hepatitis.

On the other hand, many patients have few or no symptoms with hepatitis C and infection with this virus often leads to chronic hepatitis (see below).

How infectious is hepatitis?

Perhaps the most frustrating thing about hepatitis is that a patient is most infective before he realizes that he has the disease! With hepatitis A, the most infectious period is the 2–3 days of flu-like symptoms before the jaundice develops. Once the jaundice appears there is comparatively little risk of infection and this little risk rapidly disappears over the next few days. In this type of hepatitis, precautions (see below) after the first few days are almost certainly unnecessary.

Hepatitis B on the other hand is a good deal more infectious. Most people with acute hepatitis B are capable of transmitting the disease 2 or 3 *weeks* before they become ill, and once the jaundice develops they may remain infectious for several months or even longer. Hepatitis B is diagnosed by a blood test in which the hepatitis B antigen is detected. This is not the actual virus itself, but it is a substance associated with the virus. The disappearance of the hepatitis B antigen from the blood is a clear indicator that the patient is no longer infectious. This is a relatively simple blood test which is widely available and can be repeated at regular intervals to check for continuing infectivity. In some people, however, the hepatitis B antigen never disappears from the blood, but this does not necessarily mean that they continue to be infectious. There is, in addition, a further antigen called 'e antigen'. If present in the blood it denotes infectivity but when it disappears and is replaced by e *antibody*, then the patient is no longer infectious, even if hepatitis B antigen continues to be present. These tests are the only way of detecting continuing infectiveness of hepatitis B. Patients may continue to be infective long after they have apparently recovered completely from the hepatitis. At the moment we know very little about how infectious viruses C and D are.

What precautions are necessary to preventing infecting others?

One of the major concerns that people with hepatitis have, is that they may pass the infection on to others. It is for this reason that it is very useful to know the exact nature of the hepatitis since, as mentioned above, hepatitis A is much less infectious than hepatitis B. Until the results of these tests are known, precautions to minimize the risk of infection should be taken. Kissing and other very intimate contact should be avoided. The patient should have his own cutlery and plates, which should be kept apart and washed with great care since the temperature of tap-water (or even a modern dishwasher) is not high enough to kill the hepatitis viruses. If the patient is at home, particular care must be taken in dealing with blood, faeces, or urine as all these are potentially infectious. It is rarely necessary for people with hepatitis to go into hospital, although in several countries in the world admission is compulsory for public health reasons. They are usually nursed in an infectious disease ward, where they are isolated from other patients, and visiting is restricted. These strict precautions have to be taken in hospital because of the risk of transmitting the infection to patients with other illnesses. Patients who are left at home, however, do not need to be treated with such strict isolation, but it is advisable that visiting by friends should be restricted. Ill people, the elderly, and young children should be excluded, but there is no reason why fit people should not be allowed to visit. There is very little risk, if close contact is avoided and if simple common sense is applied.

What treatment is available for hepatitis?

In common with most other virus infections there is no specific treatment for acute virus hepatitis. Fortunately, however, the large majority of patients make a complete and uneventful recovery quite quickly. Treatment is available for patients who develop certain complications of acute hepatitis and also for many of those who develop chronic hepatitis. These will be discussed later in the chapter.

Do I have to go to bed?

Until quite recently it was widely accepted that bed-rest was necessary if people were to recover from hepatitis. This is no longer thought to be necessary. If you feel ill (as usually happens in the 2 or 3 days when the flu-like symptoms are present) then you should go to bed; once, however, you feel better there is no reason why you should not get up and do as little, or as much, as you feel physically capable of. There is no evidence at all that continued activity delays the rate of recovery or increases the risk of any relapse.

Is there any special diet?

It is widely believed that patients with hepatitis should go on a strictly fat-free diet. There are two reasons why this erroneous belief is held. Many of the symptoms of hepatitis are similar to those of gallstones, and since a low-fat diet helps to relieve these (see Chapter 3) people make the mistake of thinking that such a diet will also help hepatitis. Second, during the early stages of hepatitis most people lose their appetite for all food, but particularly for fried and fatty food. This usually only lasts for a few days but in some people it may persist for longer.

Current practice is to encourage people to eat whatever they like as long as it does not upset them. During the first few days of the illness, when patients do not feel like eating any food at all, it is important to encourage an adequate intake of fluids, so that they do not become dehydrated. Tea, orange squash, and drinks containing sugar are to be preferred to water, since they help to keep up energy levels. Once the initial symptoms have improved, and the appetite is returning, patients should eat whatever takes their fancy. It is probably foolish to try large cream cakes and thick rashers of fatty bacon right from the start, but equally it is unnecessary to eat toast with no butter for weeks on end.

With a little bit of common sense, one can very quickly find out what food is tolerable. Even if certain foods cannot be tolerated they will probably do no more than cause a little indigestion. They will not in any way harm the liver. People with acute viral hepatitis lose as much as 6 or 7 kg (1 stone) in weight during the first week of the illness. Fat is a very good source of calories and so the better the diet, the quicker this lost weight can be regained.

What about alcohol?

Alcohol can harm the liver (see Chapter 6), and so it seems sensible to avoid a further insult if it is already damaged by a virus. For this reason we usually advise people to abstain from alcohol while they have hepatitis, and also for some period afterwards to allow the liver a full chance of recovery. Although we do not know how long this takes, the traditional advice has been to avoid alcohol for 6 or even 12 months. There is little evidence to support this view, and it is probable that this advice is widely ignored by teenagers and young adults! A more realistic policy is for doctors to recommend complete abstinence from alcohol until patients have recovered fully from their illness, and thereafter to allow them cautiously to resume drinking. The patient who has too many drinks for the first time may regret it, as he becomes acutely unwell with sickness and pain over the liver. This is an unpleasant reaction that warns him that he has been too impatient, but it does not, in itself, do any long-lasting harm.

It is important to remember that alcohol itself can cause hepatitis; people with alcoholic hepatitis will recover if they abstain from alcohol, but if they resume alcohol consumption the damage to the liver is likely to continue.

Should I continue taking my tablets?

Inevitably, some people who develop hepatitis may be taking medicines for other quite unrelated conditions, and they may be concerned about the possible effect that the liver damage may have on the action of these drugs (see Chapter 7). In most cases there is no need to change the dose of tablets at all, but there are some important exceptions such as anticoagulant tablets. Patients with hepatitis are in general advised to consult their doctor about the dose of any drugs.

Young women are as susceptible to virus hepatitis as the rest of the population, and many of these women may be taking the oral contraceptive pill. While they are often advised to stop taking 'the pill', there is little evidence to suggest that its continued use does any harm. (Occasionally the pill itself may cause jaundice within the first two or three months.)

It is important to remember that many drugs can actually cause hepatitis (see Chapter 7). Clearly, if the hepatitis is caused by a drug, and not by a virus, it will not get better until the drug is stopped.

When can I return to work?

The short and simple answer to this question is that you can return to work (or school) as soon as you feel well enough to do so. It is very difficult to predict when this will be, since some people recover much more rapidly than others. Some patients may take many months to regain their energy. Today, however, we encourage people to return to work much earlier than we used to, and there is no evidence that they come to any harm by so doing. There are however one or two qualifications to this advice.

It should be remembered that viral hepatitis, particularly hepatitis B and non-A non-B, are infectious conditions, so if you are involved with a job which carries a high risk of infecting other people, you may need to avoid these aspects of your occupation even if you are feeling well. This applies particularly to staff in the medical and catering professions. When in doubt advice should be sought from the general practitioner and the medical officer. Many people recovering from hepatitis feel well, but remain quite markedly jaundiced. Even though they may not be infective at the time, their appearance may cause considerable alarm and concern among their colleagues at work. Under such circumstances it is probably advisable to stay at home until the jaundice is less apparent.

Physical exercise will do no harm to patients recovering from hepatitis, but many find that they tire easily (see 'post-hepatitis blues', p. 50).

Can I get hepatitis more than once?

Recovery from a bout of hepatitis A gives one lifelong immunity against that particular virus. It does not, however, offer any immunity against the other viruses which can cause hepatitis and the same applies to hepatitis B. Some people, therefore, may be unlucky enough (or careless enough!) to acquire hepatitis on several occasions due to different viruses.

Occasionally, however, people recovering from a particular bout of hepatitis suffer a relapse. This is usually due to further symptoms, because the virus has not been completely eliminated from the body, rather than a second infection from another source.

Can I be a blood donor after getting hepatitis?

In the past anyone who had virus hepatitis was automatically excluded from becoming a blood donor. This is no longer so, although prospective blood donors are still asked whether they have ever had hepatitis. The blood of all donors, whether or not they have had hepatitis, is checked for hepatitis B and C, beforehand. Many blood transfusion centres do not accept donors who have had hepatitis within the previous 12 months, but do accept them thereafter, as long as there is no evidence of hepatitis B or C.

Can I get hepatitis by being a blood donor?

The answer to this is definitely 'no'. Some people are discovered to have hepatitis B or C when they go to give blood, because the blood is screened for these viruses every time they attend as donors. It does not mean that they acquired the hepatitis when giving blood. Although hepatitis can be transmitted by needles, this only happens if the needles used are infected or contaminated. All needles used in blood transfusion centres are very carefully sterilized, so there is absolutely no risk of transmitting hepatitis at these centres, in developed countries.

What complications can arise with acute hepatitis?

Ninety-nine per cent of people who acquire acute hepatitis recover very quickly and uneventfully from their illness, but very occasionally complications can arise.

Prolonged jaundice In some people the jaundice persists and becomes more severe over several weeks. The jaundice itself is not a serious condition, except that it is often accompanied by troublesome itching. Patients with persisting deep jaundice are usually referred to hospital for further investigation, in order to exclude other possible causes of jaundice, such as gallstones and cancer of the pancreas (see Chapter 4).

Very severe hepatitis Sometimes the symptoms of acute hepatitis are unusually severe, with patients feeling extremely ill and experiencing severe pain or copious vomiting. Such people require urgent admission to hospital, but they usually settle down after a few days of treatment, which involves pain-killers and intravenous fluids to correct their dehydration from the vomiting. Very rarely, the liver damage can be so severe that almost the entire liver is destroyed. The patients then develop liver failure, become unconscious, and usually die. This fulminating type of hepatitis pursues a very rapid course, from normal health to death, within little more than a week. Hepatitis A, B, or non-A, non-B, can all cause this rare complication.

Chronic hepatitis An intermediate situation, between complete recovery and acute fulminant hepatitis, exists. Some patients improve after the acute illness but do not return completely to normal health. They often feel generally unwell with tiredness, pain, and possibly intermittent jaundice, dark urine, pale stools, and itching. Special investigations show that there is continuing inflammation of the liver. This may resolve over a few months, but, in some people, it may continue and lead to cirrhosis (Chapter 8). We do not know why chronic hepatitis develops, although it is undoubtedly related to the persistence of the virus within the liver. Special tests including liver biopsy (Chapter 2), are required to assess the severity of liver damage, and this will require referral to hospital. Both hepatitis B and C can result in chronic liver damage, whereas hepatitis A does not.

Is there any treatment for chronic hepatitis? Several new drugs have been developed for the treatment of hepatitis B, C, and D, the most promising of which is called interferon. This drug is expensive and has to be given by injection 3 times a week for at least 3 or 4 months, and in some cases for many years. It is only effective in some people and at present should only be used under supervision from specialists who have experience with it, since interferon may cause a number of side-effects. Other new drugs are being developed and it is likely that, in the next few years, the treatment for chronic viral hepatitis will improve substantially.

'Post-hepatitis blues' It has already been mentioned that the rate of recovery from hepatitis varies considerably from one person to another. Some appear to recover very satisfactorily in that the inflammation of the liver settles down quickly, and jaundice and all other signs resolve. Despite this, they are left with a number of rather vague indefinable symptoms which cause a great deal of distress. They tire very easily, are unable to concentrate, and become easily depressed and irritable. This is often aggravated by those around them implying that they are malingering, and hinting that it is 'really time that you got back to work'. These symptoms indicate a depression which often accompanies all kinds of virus illnesses. There is unfortunately no real treatment for this 'post-hepatitis blues' except patience and reassurance. These symptoms invariably resolve with time although this may take several months. This condition has to be distinguished from chronic hepatitis by means of special tests.

Is there a relation between AIDS and hepatitis?

Although hepatitis is commonly found in groups of people in whom there is a high incidence of AIDS (homosexuals, drug addicts, haemophiliacs), the only common factor is that both diseases are transmitted by blood. Sufferers from AIDS develop all sorts of unusual infections, but hepatitis is not one of these.

Can one be vaccinated against hepatitis?

An injection of gamma globulin gives good, if not complete, protection against hepatitis A for a few months. It is generally recommended to travellers who are going to be living in fairly primitive conditions, in the developing parts of the world, where hepatitis A is very frequent. Recently a specific vaccine for hepatitis A has become available. It appears to be safe but as yet we have little experience of its effectiveness.

There are also two different methods of protecting against hepatitis B. A course of three injections of hepatitis B vaccine provides very effective protection for many years, but these injections have to be given over a period of 6 months, and there is no significant protection for the first 3 months. The main use of this vaccine is to protect those at high risk of developing hepatitis, such as certain groups of doctors and nurses, and also people working in parts of the world where hepatitis B is very common. It is also useful to protect the spouses, and other close relatives, of people who are hepatitis-B-antigen and e-antigen positive, and are likely to remain so for a long time. Although this vaccine is readily available, its present high cost limits its widespread use. In addition to this active form of vaccination, there is also available another type of vaccine in the form of a specific hepatitis B globulin. This is prepared from the blood of patients who have just recovered from hepatitis B infections, and who no longer have hepatitis B antigen, but have large amounts of specific hepatitis B *antibody* (antibodies are the molecules which neutralize antigens). This hepatitis B globulin is specifically effective against hepatitis B virus, and not against other viruses. This is effective for several months, and it is given to people who have recently been accidentally contaminated by hepatitis-B-positive infected blood, such as surgeons who have cut themselves while operating, or nurses who have been bitten or scratched by a violent patient.

No vaccine has yet been developed against hepatitis C or E. Similarly there is no vaccine against hepatitis D, but for reasons mentioned on p. 40 protection against hepatitis B will automatically protect against hepatitis D.

6 *Alcohol and the liver*

Excessive alcohol consumption is a well recognized cause of cirrhosis of the liver, but the view that cirrhosis occurs only in alcoholics is far from the truth. Alcohol accounts for about 60 per cent of cases of cirrhosis in Britain; the proportion is greater in countries where alcohol consumption is much higher, such as France, Australia, and the USA. It is lower in those parts of the world where little alcohol is consumed or where other causes of cirrhosis (see Chapter 8) are common, such as India and the Middle East. No race is immune from alcoholic liver damage. All forms of alcohol can cause cirrhosis (Fig. 13).

It is important to realize that liver damage is not inevitably associated with either alcoholism or excessive drinking. These terms have already been discussed in greater detail in another book in this series, but a brief and widely accepted definition of an alcoholic is 'one who is dependent upon alcohol and in whom alcohol causes anti-social behaviour'. In some cases the amount of alcohol required to induce this is quite small, and the majority of alcoholics do not develop progressive alcoholic liver disease.

Excessive drinking is very difficult to define, since the definition of excess depends very much upon what is the accepted norm in any given society. Thus, in a strongly religious puritanical society, anything more than the weekly sip of communion wine may be regarded as excessive, whereas, in wine-growing areas of France, or in coal-mining communities of Western Europe, a daily bottle of wine or eight pints of beer are considered to be 'normal'. The phrase 'an excessive drinker is someone who drinks more than I do' is not quite as flippant as it sounds. Since alcohol is such an integral part of our social life, it is society which tends to set the standards of normality.

It appears that certain people are unduly sensitive to alcoholic liver damage, and are liable to develop cirrhosis

Fig. 13 All forms of alcohol can cause cirrhosis.

with quite small amounts of alcohol. Only a minority of alcoholics and people who drink excessive quantities of alcohol for a sustained period of time appear to develop cirrhosis. The reason why some people are apparently immune to liver damage is unknown. The risk of developing alcoholic cirrhosis among sustained heavy drinkers can be compared to Russian roulette; the odds are about 1 : 6.

Many other organs in the body are also susceptible to damage by alcohol, independently of damage to the liver (see below).

To summarize: alcoholism, excessive drinking, and alcohol-induced liver damage are very distinct problems which are not necessarily related to one another.

How much alcohol is harmful?

It is impossible to give more than broad guidelines in answer to this question. We have already implied that some individuals are more susceptible to alcohol than others, so what is apparently quite safe for one person is potentially dangerous for another. The toxic element in all alcoholic beverages is the alcohol itself (or, to use its proper chemical name, 'ethyl alcohol'). It is ethyl alcohol alone which in the short term produces the acute effects of alcohol intoxication (i.e. drunkenness), and in the long term produces liver damage. In general terms, the more alcohol consumed, the more likely the chances of developing damage, regardless of whether it is drunk as gin, wine, beer, or champagne. Other substances, called congeners, exist in alcoholic beverages. These may produce a whole variety of unwanted side-effects such as headaches, flushing, and sickness, but these do not cause liver damage. Figure 14 shows the comparative contents of a variety of commonly consumed alcoholic beverages. In simple terms, a pint of beer is equivalent to a double measure of spirits, while a bottle of table wine is equivalent to five pints of beer. This figure is only a very rough guide and it is worth noting a number of important points with regard to alcohol content. Many alcoholic drinks vary widely in their alcohol content according to particular type. The figure quoted for beer refers to draught beer in Southern England; bottled beers and those brewed in other regions tend to have a much higher alcohol content, while most home-brew enthusiasts tend to produce a beer with twice or even three times the amount of alcohol. Contrary to popular belief, cider usually contains more alcohol than beer and here again the draught cider produced in Western England is particularly potent. People invariably under-estimate the amount of alcohol they consume, whether or not they are alcoholics. This is especially true when drinking at home and pouring one's own measures. A single measure of whisky in your own living-room is usually equivalent to a double, or even treble, purchased in the pub. However little you drink, you are almost certainly drinking more than you think you are!

It is generally agreed that the regular daily consumption of 60 g of alcohol, equivalent to three and a half pints of beer, or a quarter bottle of spirits, places the average man at risk of developing serious liver damage. For women the figure is a good deal lower, and while a daily consumption of as little

Fig. 14 Approximate alcohol content (in grams) of some popular drinks.

as 20 g of alcohol has been quoted as being dangerous, the generally accepted safe level is probably in the region of 30–35 g (two pints of beer or two double gins). It should be emphasized that these figures represent an average daily consumption which must be sustained over many months or years. An evening out drinking half a bottle of whisky once every few months will not do any serious harm to the liver,

Fig. 15 'I only have one drink a day, Charles!'

even though your head may suffer the next morning! The reason why women are more susceptible than men to alcohol is unknown. It may be related to the fact that women are physically smaller than men, so any given amount of alcohol may have a greater effect on a smaller body. Large overweight women should not feel that they can drink greater amounts of alcohol safely however, since this is, at best, only part of the explanation. It has also been found that an enzyme in the stomach called alcohol dehydrogenase breaks down alcohol more slowly in women than in men, so the effects of alcohol are likely to last for longer. Hormonal differences between the sexes may also be important.

How does alcohol damage the liver?

Serious damage to the liver requires a regular alcohol intake which is sustained for at least one or two years. A single night or weekend of drinking may induce acute ill-effects on the liver, but this will not result in any long-term damage. The main acute effect of alcohol is to cause fat to be deposited within the liver. This fat usually stays there for a few days and then disappears if the individual abstains from alcohol, but if drinking is sustained the fat persists. In most people the fatty changes persist to a greater or lesser extent throughout life with no significant progression, but in susceptible individuals progressive changes occur which lead to cirrhosis. Some heavy drinkers may develop an acute illness with jaundice which resembles virus hepatitis (see Chapter 5), and it is only by means of sophisticated tests that these two conditions can be distinguished. If the hepatitis is due to a virus, the patient may safely resume moderate alcohol consumption when recovered, but all individuals with alcoholic hepatitis must abstain completely if they are to avoid cirrhosis. The reason why some individuals develop alcoholic fatty liver, while others develop hepatitis is unknown.

What are units of alcohol? Most people have difficulty in calculating alcohol consumption in grams, since different drinks have different amounts of alcohol. If you look carefully at Fig. 14 (p. 56) you will see that a single measure of whisky, a glass of sherry, half a pint of beer, and a glass of table wine each have the same quantity of alcohol—namely 8–10 g. This quantity of alcohol is defined as a 'unit' and it enables people to equate different drinks so that they can use it as a rough and ready guide to alcohol consumption. Thus, one pint of beer is two units, which is equivalent to a double measure of spirits. A bottle of table wine is about 10 units, while a bottle of gin or whisky is about 30 units. Using this scale the maximum recommended alcohol intake which is considered safe for a man is 21 units per week (3 units per day), and 14 units per week (2 units per day) for a woman. These figures are much lower than the 60 g (8 units per day) and 30 g (4 units per day) quoted for men and women on p. 55, but these lower values refer to safe levels for *all* individuals in order to prevent *any* kind of alcohol-related damage, rather than for liver damage in some people. Although the concept of units of alcohol is simple and easy to remember, it should be stressed that it is only a very approximate guide; it does not, for example, take into account that some beers or wines are stronger than others. Neither does it acknowledge that measures of spirits vary from country to country, or that measures poured at home are much larger than those served in a bar!

What are the symptoms of alcoholic liver damage?

Unfortunately most people with progressive liver damage have few symptoms until the disease is far advanced, and their first symptoms are those of the complications of cirrhosis (see Chapter 9). These include distension of the abdomen due to fluid (ascites), bleeding from oesophageal varices, and disturbances in mental function (encephalopathy). By this stage however, extensive and largely irreversible liver damage has already occurred.

Earlier symptoms of liver damage tend to be rather minor and non-specific. Despite the inflammation and damage which is taking place in the liver, few people seem to develop pain. This is because the inside of the liver is relatively free of nerve fibres. Even when pain does occur it is rarely over the liver, but tends to be all over the abdomen. It is usually mild and may readily be mistaken for indigestion. Diarrhoea is another common symptom, but this, like abdominal pain, is very non-specific and may be due to many other causes. Neither diarrhoea nor abdominal pain distinguish between fatty liver and more serious liver damage.

Jaundice, which is often accompanied by pale motions and dark urine, indicates liver damage, and anyone with these symptoms should seek medical advice. As mentioned earlier, these are features of viral hepatitis, but in any person who has been drinking any significant amount of alcohol for any length of time the possibility of alcoholic hepatitis must be considered.

Liver disease and gallstones

Does diet affect alcoholic liver damage?

The effect of diet on alcoholic liver damage has been a very controversial topic for many years. Long-term heavy drinkers tend to substitute alcohol for food as their major source of calories and, since alcohol alone lacks most of the important nutritious components of a healthy diet, these people tend to become thin, wasted, and malnourished. This, however, occurs irrespective of whether they have liver damage or not. Many patients with severe alcoholic liver damage first appear in this physical condition and in addition they have lost their appetite. Such patients may be desperately ill and one of the earliest and most hopeful signs of improvement is the return of appetite. Patients who die from liver failure tend to be those who do not regain their appetite, and do not manage to eat sufficient calories to sustain adequate health. Despite these observations, however, people can develop severe liver damage while continuing to eat normally.

Patients with malnutrition, or evidence of vitamin deficiencies, should be treated by an appropriate diet and vitamin supplements. For others whose appetite is good and who are well nourished, a healthy balanced diet with a good calorie intake is to be recommended. In a few patients with cirrhosis and encephalopathy (see Chapter 9) it may be necessary to restrict protein. The best diet for a patient with alcoholic liver damage is a balanced high calorie diet which contains everything except alcohol.

Is ability to drink related to liver damage?

Some people are able to 'hold their liquor' far better than others, but this does not indicate any susceptibility or resistance to the effects of alcoholic liver damage. Getting drunk, for many people, may act as a protective mechanism against excess alcohol consumption because it limits any further intake. Alcoholics often find that as their problem gets worse and worse, they are able to drink less and less before becoming incapable, but this does not mean that they are developing liver damage. The metabolism of alcohol by the body does not appear to alter significantly as liver damage progresses, at least not until the damage is extremely far advanced. It is important, therefore, to realize that cirrhosis of the liver cannot be taken as a mitigating circumstance when considering blood alcohol levels in relation to a driving offence.

Habitual drinkers frequently become dependent upon alcohol and suffer withdrawal symptoms (delirium tremens or DTs) if they suddenly stop drinking. Only some patients develop DTs, but these symptoms are quite unrelated to liver damage.

Is it necessary to abstain for life if one has alcoholic liver damage?

This depends very much upon the type of liver damage. Where there is alcoholic hepatitis or cirrhosis the answer is quite clear cut: life-long abstention from alcohol is the only way to ensure that the liver damage will not progress. People who have already developed alcoholic hepatitis or cirrhosis have livers which are undoubtedly sensitive to alcohol, and continued drinking will result in further damage to the liver, even if the rate and pattern of alcohol consumption is altered. Reducing alcohol consumption will reduce the rate of progression of the liver damage, but it is most unlikely to prevent its progress. Patients with symptoms due to alcoholic liver damage (such as pain, diarrhoea, or jaundice) may well find that drinking less alcohol leads to improvement or even disappearance of the symptoms. This, however, does not mean that the liver damage is no longer taking place. As we have pointed out earlier, much damage to the liver occurs in the absence of any symptoms.

It is much more difficult to answer this question with regard to fatty liver. People who have been drinking for many years and who have mild degrees of fatty liver will not progress to cirrhosis, and any reduction in their alcohol intake is likely to result in improvement of their symptoms, with little or no substantial risk of developing progressive liver damage. On the other hand, people who have been drinking for a relatively short time, perhaps 5–10 years or less, and those who have very large amounts of fat in the liver, may well be at much greater risk of developing cirrhosis if they continue to drink. Such people are advised to have regular follow-up and medical supervision.

How can alcoholic liver damage be detected?

The individual who is drinking significant amounts of alcohol on a regular basis cannot rely upon the appearance of symptoms as a reliable indicator of liver damage. Anyone concerned about the possibility of having liver damage should seek medical advice. Often a full and careful medical examination may show signs of liver inflammation and damage, but unless the damage is very far advanced it will be difficult, if not impossible, to distinguish between the relatively benign fatty liver and the more serious alcoholic hepatitis and cirrhosis. There are a number of blood tests available which can show signs of liver inflammation and damage, but once again these are unlikely to discriminate between different types of liver damage, unless it is far advanced. The only reliable way of distinguishing between the different types of liver damage is by liver biopsy (see Chapter 2).

Unfortunately liver biopsy requires admission to hospital and is a relatively expensive investigation, which is not without some degree of risk, albeit small. Not surprisingly, many patients and doctors are reluctant to pursue this course. An alternative approach is for the patient to stop drinking altogether for a period of a few weeks or so, and then to see whether the abnormalities which were initially detected have resolved completely. If they persist, despite this period of abstinence, then the indications for further investigation are very strong. We urgently need simple tests to enable us to distinguish between the different types of liver disease, but as yet none is available.

Is alcoholic liver damage reversible?

This depends entirely upon the nature of the liver damage and the extent to which it has progressed. On the one hand, fatty liver is completely reversible within a few weeks of stopping drinking and, similarly, alcoholic hepatitis can resolve completely, although this usually requires a longer period of abstinence. More severe degrees of liver damage may settle down, but will leave the liver with a certain amount of scarring which need not necessarily cause any long-term harm to the patient. Once cirrhosis has developed, some degree of permanent damage will remain even if the individual never drinks another drop of alcohol. Nevertheless there is still a great deal of recovery and improvement possible in any patient with established cirrhosis. This is because the cirrhosis is usually accompanied by a good deal of liver inflammation, which can settle down once alcohol is withdrawn, although it often takes many weeks or even months for this to happen. Unfortunately, some people seek medical advice for the first time only when the liver inflammation is very active and well advanced, and by then it may be too late for any reversal to occur even if they do stop drinking. Broadly speaking, however, it is never too late to stop drinking.

Patients with complications of alcoholic cirrhosis (see Chapter 9), who continue to drink substantially, have a life-expectancy which is comparable with many forms of cancer.

Is alcoholic cirrhosis hereditary?

There is no good evidence that alcoholic cirrhosis runs in families although there is a suggestion that alcoholism does. Even this is somewhat controversial, since there can be little doubt that environment plays a significant role in drinking problems. It is easy to understand how anyone brought up in a situation of domestic strife, due to one or both alcoholic parents, may ultimately become an alcoholic.

Can alcohol damage other parts of the body?

Few organs in the body are immune from the effects of alcohol, but such effects occur independently of liver damage. Most of these are outside the scope of this book but a few will be mentioned here since they may be relevant to alcoholic liver disease.

Alcohol may damage the brain and the nervous system in a variety of ways. It is well recognized that habitually heavy drinkers develop dementia. They tend to become forgetful and lose the ability to concentrate. The brains of such people are much smaller than normal. Similar changes can also be seen in people with advanced liver disease who develop encephalopathy (see Chapter 9). Here, the brain damage is not due to a direct effect of alcohol on the brain, but occurs as a consequence of the severe liver damage. In some people, of course, both processes may be taking place at the same time.

As mentioned earlier, many heavy drinkers develop abdominal pain and only in some cases is this due to liver inflammation. Inflammation of the stomach (gastritis) may cause pain, sickness, and vomiting. The pancreas is another organ which is sensitive to the effects of alcohol and, when inflamed, this can cause severe pain. The common bile-duct passes through the middle of the head of the pancreas (see p. 4), and so inflammation of this organ may lead to compression of the duct and consequent jaundice. Thus, although pancreatitis may be completely unrelated to alcoholic liver damage, the symptoms of both conditions may be very similar. Alcohol may also have marked effects upon the reproductive organs. The mechanism whereby this occurs is poorly understood, but probably it can affect the organs either directly or indirectly through damage to the liver, which is a well recognized site of sex hormone production (see p. 8). The effects in men are to produce growth of the breasts (gynaecomastia), shrinkage of the testis, impotence, and loss of body hair and libido. Cessation of menstrual periods and infertility are the major symptoms in women. Even if women who drink heavily do get pregnant, they have a higher rate of miscarriage. Should the pregnancy proceed to term the baby is likely to be smaller than average and to have a risk of being born with various congenital abnormalities of the face and jaw (this is usually referred to as the fetal alcohol syndrome).

Giving up drinking: What help is available?

Many alcoholics and heavy drinkers who ignore the pleas of friends and relatives to stop drinking take a different attitude when confronted with clear evidence of damage to their health. Nevertheless, for many this is a very difficult action to take since it frequently means changing the habits of a lifetime, and often requires active help and support. The general practitioner should be the first line of support for professional help and advice. He or she is usually able to offer moral and physical support, suitable treatment for overcoming the problems of alcohol withdrawal, and also advice on appropriate organizations that may be able to help further. Many psychiatrists have a special interest in alcohol problems, and may play a valuable role with patients whose drink problem is associated with a personality disorder; their effectiveness is often limited, since many patients actively resent the suggestion that they might have some psychiatric disorder. In addition there is a growing number of non-medical organizations which have been formed to support those with an alcohol problem. Among the best known are the following.

Alcoholics Anonymous This is a self-help organization with branches in many countries around the world, in which reforming and reformed alcoholics keep closely in touch with one another for mutual support.

Al Anon This organization exists to help the families of alcoholics who frequently may suffer considerably in both mental and physical terms. The addresses of these organizations are usually available in the local telephone directory or Citizens Advice Bureau.

Al Ateen Children are all too often the victims of alcohol problems affecting one or both parents; this voluntary organization exists to support them.

Alcohol advisory services and councils on alcoholism An increasing number of cities throughout Great Britain have local advice services. These bodies rely on voluntary support as well as funding from local authorities, health authorities, and other bodies. Although run by a limited number of paid staff, they include many professionals among their volunteer counsellors. Their work covers individual counselling, group work with alcoholics and their families, and day-to-day support during the recovery period. These organizations work closely with both the medical profession and the social services.

Despite all these organizations, the most valuable support for any person trying to give up alcohol must come from their closest family and friends, who should be involved at the earliest possible stage in any treatment, because, without their active understanding and support, the chances of persuading an individual to change his life-style *and to maintain this change* are indeed remote.

7 *Other liver poisons*

Since almost everything that we eat and drink finds its way to the liver before it is broken down, it is hardly surprising that many substances which are poisonous to the body act by causing damage to the liver. This chapter will deal with the harmful effects on the liver which can be produced by drugs and toxins. By 'drugs' we mean all substances that are taken for medicinal purposes and not just narcotics such as cocaine or heroin. Drug addicts commonly acquire hepatitis; however this is not due to the harmful effect of the narcotics they are using, but to the transmission of virus hepatitis by infected blood in shared syringes and needles (see Chapter 5).

In general terms we recognize that there are two different types of drug effects: allergic reactions and toxic reactions.

Allergic reactions

For reasons which are generally unknown a small minority of individuals are allergic to certain chemicals. The most common form of allergy is a skin rash. Five to ten per cent of all individuals develop a rash when taking penicillin. Other people develop drug allergic reactions which affect the liver. These reactions usually occur several days after the first exposure to the chemical, although sometimes the time interval may be longer. They are not dose-related, that is, the reactions may sometimes develop after only one or two tablets and they are generally completely unpredictable. These reactions are often described as being 'idiosyncratic reactions' as the patient has a particular idiosyncrasy to a specific group of substances.

The nature of the liver damage

Two distinct types of liver damage may be produced by
allergic drug reactions involving the liver. These are termed
'hepatitic' and 'cholestatic'.

Hepatitis-like reactions   In this type of reaction the patient
develops an illness which is completely indistinguishable
from viral hepatitis. The signs and symptoms of this are
described in Chapter 6 and it is often very difficult to distinguish between drug and virus hepatitis. Nevertheless, it
is extremely important to do so, since in drug hepatitis the
liver damage will only improve once the offending agent is
stopped. Hundreds of drugs have been reported to cause
hepatitic reactions. Such reactions are often very difficult to
identify, since the majority occur in acutely ill people in
hospital, who may well be receiving a number of different
drugs over a short period of time. The time interval between
receiving the drug and developing the hepatitis is variable,
and may be anything from a few days to several weeks. One
particularly important drug in this group is halothane. This
is a widely used good anaesthetic agent, which is very safe,
but occasionally patients may develop hepatitis after
repeated anaesthetics. An individual who has had hepatitis
related to halothane must not receive this anaesthetic again
at any time. Although this will be recorded in the hospital
notes, the patient should also be aware of the problem in
case of admission to another hospital.

Cholestatic reactions   This is a term which is applied to
patients who develop jaundice, without any inflammation of
the liver, secondary to drugs. It is as though the liver
suddenly becomes unable to excrete bilirubin into the bile
and so the bilirubin accumulates within the blood. This is
analogous to the situation which occurs with obstructive
jaundice due to gallstones, or cancer of the pancreas (see
p. 35), except that there is no obvious obstruction to the main
bile-ducts. This too can be produced by a large variety of
drugs but the oral contraceptive pill is the best-known agent
for producing this kind of reaction. Chlorpromazine, which

is a commonly used tranquillizer, produces this reaction in about one in fifty people. Cholestasis often makes the patient feel very ill with considerable itching. The jaundice resolves once the drug is stopped but this may take several weeks or even months.

Does a history of allergy to one drug increase the chances of developing allergic reactions to other drugs?

Once a patient has developed an allergic reaction to a given drug it is likely that they will develop a reaction to any drug which has a similar chemical formula. Drugs usually belong to groups of well-defined chemicals and most doctors or chemists will readily be able to identify such compounds. Differences depend upon the chemical structure of the compound rather than upon its mode of action. There is very little danger of cross-reaction between widely differing substances.

Why is it important to recognize drug-associated liver damage?

There are three main reasons for this:

1. If a drug has caused liver damage, then the damage is likely to continue if the patient continues to take the drug. The sooner it is stopped the quicker the liver will recover.
2. Drug-induced liver damage is unlikely to respond to any other kind of treatment apart from withdrawal of the drug.
3. Once a certain drug has caused liver damage, this drug (as well as chemically similar compounds) must be avoided for life. Re-exposure to this drug is likely to cause the same symptoms to reappear, and there is a considerable risk that on a second or third exposure the symptoms will be appreciably more severe.

*Important practical points for patients who have had
drug hepatitis*

Anyone who has experienced drug-associated liver damage
should take great care to acquaint themselves with the exact
nature of the drugs which have been implicated. They
should do everything they can to ensure that they are not
unwittingly given the same drug again. They should ensure
that their general practitioner (and indeed any other medical
practitioner who attends to them) is alerted to the relevant
medical history. One very real fear and danger is that an
individual may be involved in a car accident or some other
event in which they are brought unconscious into hospital
and may in the course of emergency treatment receive
potentially dangerous drugs. This danger can be minimized
if the individual carries a record of the drug allergy in their
wallet or purse. Small, inexpensive, 'medic-alert' bracelets
can be bought and worn on the wrist to alert medical
attendants.

Toxic reactions

While some drugs cause allergic reactions, there are many
substances which will cause predictable liver damage if
ingested in sufficient amounts. All people are susceptible,
and the larger the amount of the substance ingested the
greater the damage. The most commonly encountered
example of this in Britain, today, is paracetamol. This is a
very effective pain-killer which is remarkably safe when
taken in doses of 2 or 3 tablets at a time, up to 4 times a day.
Unfortunately, it is also widely used for suicide attempts,
since it is readily available over the counter in any pharmacy.
Twenty or thirty tablets will cause serious liver damage,
while larger overdoses are likely to be fatal unless specific
treatment is given. People who take paracetamol in large
quantities often feel ill within the first 24 hours, but the really
serious liver damage usually takes 3 or 4 days to develop.
Specific antidotes to paracetamol are available and are highly
effective as long as they are given as soon as possible after

the overdose. They certainly work if they are given within 15 hours of taking the tablets, but it is doubtful whether they really have any effect once liver damage has developed. It is therefore essential that anybody taking an excess of paracetamol is taken to hospital at the earliest opportunity. Although the effects of paracetamol on the liver can be very serious, they are potentially completely reversible. Unless the patient dies of liver failure within a few days, full recovery of the liver can be expected with no long-term after-effects.

Many chemicals which are toxic to the liver are produced by, or used in, industrial processes. These include carbon tetrachloride, organic pesticides, and derivatives of benzene and toluene. Fortunately, the toxic nature of these chemicals is well recognized and so in most countries rigorous safety precautions are taken to minimize the risk to all involved. The threat of industrial toxins, however, is ever-present, since the development of new chemical processes in the future will almost certainly reveal new toxins which may harm the liver.

Plants may also produce poisonous chemicals that can damage the liver. Perhaps the best known of these are the poisonous toadstools of the Amanita species. These very rarely cause any problems in the United Kingdom, since poisonous toadstools are uncommon and also the British are traditionally suspicious about picking wild fungi. No such inhibitions are found on the continent of Europe however and it is not at all uncommon for whole families to be admitted to hospital suffering from the acute ill effects of liver damage due to 'mushroom' poisoning.

Can drugs and toxins produce chronic liver damage?

Although the large majority of drug reactions involving the liver occur within a week or two of starting the drug, we are increasingly recognizing that some compounds can produce mild subtle damage without any acute illness. This can lead to long-term inflammation, and even cirrhosis of the liver if the drugs are continued for a period of months or years. This applies not only to drugs which are taken for medication but also to a number of compounds used in industry.

Many industrial companies recognize this problem and recommend employees exposed to such chemicals to have regular medical examinations and blood tests. While these checks are usually not compulsory, employees are strongly advised to take advantage of these offers, even if they are feeling perfectly well.

8 Cirrhosis

The word 'cirrhosis' is derived from the Greek *'kirros'*, meaning tawny, which was the appearance noted by those who first described this condition. The most characteristic appearance of the cirrhotic liver, however, is the hard, irregular, knobbly surface made up of lumps (nodules) of liver tissue interspersed with pale fibrous material (Fig. 16*b*, p. 84).

How does cirrhosis develop?

Cirrhosis may arise as the long-term sequel to any type of inflammation. Normally, when the liver is acutely damaged, the liver cells die and the organ regenerates to its original size and shape without any scarring. This property of the liver has been recognized for centuries. In ancient Greek mythology, Prometheus was chained to a rock on Mount Caucasus where during the daytime the vulture fed on his liver, which regrew each succeeding night. In cirrhosis, however, this process of regrowth appears to go slightly awry. The healing process appears to be incomplete because scar tissue develops; this is called 'fibrosis'. In addition, the liver regenerates imperfectly, in that smooth homogeneous tissue is replaced by irregular nodules. The combination of nodular regeneration and fibrosis is called cirrhosis.

What causes cirrhosis?

Many different conditions affecting the liver can cause cirrhosis. Alcohol (Chapter 6), and virus hepatitis (Chapter 5), are far and away the commonest causes in the world, but by no means the only ones. A number of inherited conditions which are mainly seen in children are described in Chapter 11; further diseases which affect adults will be described briefly here.

Auto-immune chronic active hepatitis

During the last 30 years or so we have discovered a number of diseases which seem to be caused by some part of the body developing an allergy to itself. When this affects the liver it is given the rather cumbersome title of 'auto-immune chronic active hepatitis'. It is very similar in many ways to chronic hepatitis following infection with viruses (see Chapter 5), but special blood tests are usually able to distinguish between the two conditions. Auto-immune hepatitis is much more common in women than in men and is important to diagnose in the early stages, when it can be treated and controlled, if not cured. The treatment consists of steroids and azathioprine, which are two important drugs used in preventing liver transplant rejection (see Chapter 12). In this way the two conditions are similar—the body tries to reject the transplanted liver as a 'foreign' tissue, just as people with auto-immune disease try to reject their own tissue.

Primary biliary cirrhosis

This is another condition which is much more common in women, particularly after the age of 50. These women usually complain of itching and jaundice initially, and gradually over a number of years the liver disease progresses to cause cirrhosis and liver failure. The rate at which this disease progresses, however, is very variable and unpredictable, with patients often leading a healthy symptom-free life for many years after the diagnosis is first made. This is particularly true in the case of patients who are diagnosed by chance. An increasing number of people today undergo routine blood tests as part of health checks, and the finding of abnormal liver enzymes often leads to further investigation and a diagnosis of primary biliary cirrhosis. The cause of the disease is completely unknown, although in some people it may resemble auto-immune chronic active hepatitis. Once again special blood tests are needed to make the diagnosis in addition to liver biopsy. There is at present no treatment for this condition although a number of new drugs are being tested. If and when the disease progresses to an advanced stage, patients with primary biliary cirrhosis are usually particularly suitable candidates for liver transplantation (see Chapter 12).

Secondary biliary cirrhosis

Patients who have a blockage of the bile duct (see Chapter 4) for a long time may develop cirrhosis. This is called secondary biliary cirrhosis because it occurs as a consequence of a problem in the bile-duct rather than being a primary damage to the liver. This is most frequently due to a narrowing or stricture of the bile-duct, which occurs either as a result of gallstones or due to operations which have been carried out to remove the gallstones.

Primary sclerosing cholangitis

Some patients with inflammatory bowel disease (ulcerative colitis or Crohn's disease) may develop inflammation of the bile-duct which can lead to chronic liver damage and cirrhosis. Most people experience little or no symptoms while developing this, and it is not known why only some individuals with inflammatory bowel disease develop it. Unfortunately, sclerosing cholangitis cannot be prevented simply by removing the diseased bowel or by controlling the bowel inflammation.

Haemochromatosis

This is a disease in which abnormal amounts of iron are absorbed into the body and deposited in various organs including the liver. Iron is present in much of our food, especially meat, eggs, and cereals. In some people this condition arises as a result of alcohol abuse, since alcohol is able to increase the absorption of iron from the bowel. In many, however, there is no relation to alcohol and these patients have a predisposition to absorb too much iron. This abnormality is genetically determined, so once a patient has been identified as having haemochromatosis, other members of the family should be checked for evidence of the disease. It very rarely affects women; their normal menstrual blood loss protects them from accumulating too much iron. The condition is treatable by removing iron from the body either with drugs or by regular blood-letting.

What symptoms does cirrhosis give?

Patients with uncomplicated cirrhosis often have few complaints. We have already described the symptoms of alcoholic liver disease in Chapter 6. These include diarrhoea, abdominal pain, and jaundice, but patients may also experience the more vague symptoms of tiredness, weakness, and loss of appetite. All of these may be features of any type of cirrhosis—alcoholic or otherwise.

Why is cirrhosis serious?

Cirrhosis is a serious condition. Once developed, it is irreversible, even if the underlying cause which has led to it can be removed or eliminated. This does not mean however that treatment is hopeless, for if the inflammation within the liver can be reduced then the rate of progression of liver damage can be slowed down or even halted. If cirrhosis is unchecked, the liver will become increasingly replaced by fibrous tissue until finally there are not enough liver cells to maintain the various functions which have been described in Chapter 1. At this stage the patient dies of liver failure. Complete failure of the liver however is only one of the complications of cirrhosis, and one or more other symptoms usually develop before this stage is reached. One of the commonest problems is portal hypertension, a condition caused by an increase in pressure in the portal vein. This gives rise to ascites and oesophageal varices. In 80 per cent of cases this is due to cirrhosis, but it may occasionally be caused by thrombosis of the portal veins or the hepatic veins, as well as compression by a primary or secondary tumour

(Chapter 10). In certain parts of the world schistosomiasis (see p. 109) is a major cause of portal hypertension without cirrhosis. It is the complications, discussed in greater detail in the next chapter, which are life-threatening.

Is cirrhosis caused by a deficient diet?

Although cirrhosis can be induced in rats by feeding them with a diet which is deficient in certain amino acids, there is no evidence that dietary deficiency causes cirrhosis in humans. Many people with alcoholic cirrhosis appear to be wasted and malnourished, but this is usually because they have stopped eating food in favour of drinking alcohol. It is the alcohol excess, rather than the lack of food, which has caused the cirrhosis. Conversely, a good diet does not protect you against alcoholic cirrhosis if you drink excessively, although it may influence your general state of health and your appearance when the signs of alcoholic liver disease appear.

Should I alter my diet if I have cirrhosis?

People with cirrhosis should be advised to eat a well-balanced nutritious diet with plenty of calories. Foods which upset you should be avoided, but there is nothing to suggest that this is any more common in patients with cirrhosis than in the population as a whole. People with very advanced liver disease are often thin and malnourished, either because they are not eating enough or because the damaged liver affects their ability to absorb food. Whenever possible such patients should be encouraged to eat as much as possible, because the more food they can eat the more will usually be absorbed and the stronger they are likely to be.

Like calories, vitamins are an important part of the diet, and people who are not eating properly nor absorbing their food may become deficient in one or more vitamins. It is for this reason that patients with liver disease are often pre-scribed vitamin supplements. Vitamin deficiency does not make liver disease worse, but it may produce added complications to those of the liver damage. People whose nutritional state is satisfactory and who are eating a good balanced diet do not require vitamin supplements as they will obtain all the vitamins they require from the diet alone. Although vitamins are essential for good health, it is possible to damage both the liver and other organs in the body with excess vitamins.

There are, however, a few dietary constraints for patients with liver disease. Alcohol consumption is important regardless of the cause of the liver damage! Abstinence in those whose liver is already damaged by alcohol has already been discussed in Chapter 6. In non-alcoholic liver disease, excessive alcohol consumption should be avoided since it is foolish to expose an already damaged liver to something else which is known to be harmful to the liver. An occasional glass of beer or wine is unlikely to do any harm, but care should be taken to prevent this becoming a regular habit. Another dietary limitation concerns protein, and this only applies to people with very advanced liver disease. As described on p. 82, too much protein may affect the brain causing confusion and loss of concentration. Finally, patients with cirrhosis which is complicated by ascites (see Chapter 9) are usually advised to moderate their salt intake, since too much sodium aggravates retention of water.

9 Complications of cirrhosis

Ascites (fluid in the abdomen)

Cirrhosis of the liver leads to obstruction of the blood flowing through the portal veins, and as a result the pressure within these veins rises. This increased pressure tends to push the fluid out of the portal vein and its branches. The red blood cells are too large to be pushed out of the veins, but the remaining fluid (the plasma) seeps into the space surrounding the intestine known as the peritoneal cavity. This accumulation of fluid is known as 'ascites'.

This increase in the pressure in the portal veins is only one of the mechanisms that leads to the formation of ascites. Cirrhosis also seems to affect the function of the kidneys so that they can no longer get rid of water and salt from the body in the normal way. This leads to an accumulation of salt (or more precisely sodium) and water within the body, which accumulates in the abdomen.

What are the symptoms of ascites?

The commonest symptom of ascites is that of steadily increasing abdominal girth. The distension often occurs very slowly over a number of weeks or months and so people may not realize that there is any significant change until it is quite gross. It is not uncommon to see people whose abdomen contains as much as 8 or 10 litres (15–20 pints) of fluid. The tightening of clothing round the waist is the most obvious way to gauge any recent increase. Many heavy drinkers tolerate the abdominal distension in the belief that this is just a 'beer-drinker's paunch'. In women this gives the outward appearance of pregnancy. Some people go along to their doctor complaining of loss of appetite. This is because the distension gives them a sensation of permanent fullness, similar to that experienced after a large meal. As the abdomen gets larger and larger the pressure within it tends to impede the return of blood from the legs, and this in turn gives rise to swelling of the feet and ankles. This is called 'oedema' and patients often go to see their doctor for the first time when this swelling prevents them from putting on their shoes.

What other conditions cause abdominal swelling?

Fat and pregnancy are by far and away the commonest causes of increasing abdominal swelling. Benign ovarian cysts and cancer of the ovary are two further common conditions which can affect women. Ascites can also be produced by a number of serious diseases including cancer, very severe heart failure, and kidney disease.

How is ascites treated?

Treatment is generally aimed at stimulating the kidneys to get rid of the excess salt and water. This is done by drugs known collectively as diuretics. Since different diuretics work in different ways on the kidneys, it is quite common to use a combination of two different tablets especially when higher doses are required. There is a wide variety of different diuretics available, but frusemide and spironolactone are two of the most widely used.

Part of the problem concerns the inability of the kidney to get rid of salt. Therefore, many patients are advised to reduce the intake of salt in their diet rather than to try to maintain a complete salt-free diet which is very restrictive and unpalatable. Reduction of salt intake usually results in a smaller dose of diuretic tablets being required.

Why not drain off the fluid with a needle?

This would certainly produce relief of the tension much more rapidly than diuretic tablets, which often take several days or even weeks to work. Unfortunately there are problems associated with this. First of all the fluid tends to reaccumulate within a few days and repeated drainage may make the patient ill. Nevertheless, small amounts of fluid can be removed from time to time to relieve severe and painful distension.

Is the presence of ascites dangerous?

Ascites, when gross, is uncomfortable and unsightly, but in itself is not particularly dangerous. The danger arises if it becomes infected, a condition known as peritonitis, which can be rapidly fatal if not detected and treated. The presence of ascites in cirrhosis usually indicates that it is well advanced and that other complications are likely.

Encephalopathy

This is a complication of cirrhosis which affects the brain. Patients with encephalopathy develop a variety of different symptoms, ranging from minor lapses in memory and inability to concentrate, to slurred speech and confusional episodes which may be transient or permanent. As it progresses, patients become sleepy or even unconscious. It is not difficult to see how these late complications are seriously debilitating, but in the early stages the changes may be very subtle. The Oxford don who takes an extra five minutes over *The Times* crossword is likely to notice slight intellectual impairment well before the unskilled labourer!

What causes encephalopathy?

Encephalopathy is believed to be caused by toxic substances, such as ammonia, which are formed when protein in the diet is broken down by the action of many millions of bacteria which inhabit the intestine. The toxins are absorbed into the portal vein and normally removed from the blood by the liver; but when the liver is badly damaged, they pass through it and go to the brain where they have direct harmful effects.

How can encephalopathy be treated?

Fortunately much progress has been made in the treatment of this disabling condition in recent years. Since we know that protein is the source of these toxic substances, many of the symptoms can be improved by simply restricting dietary protein. Whereas the normal diet contains about 70 g of protein it is possible to reduce the intake to 40 g a day without any great ill-effect. Reduction of protein below this level is theoretically possible but is usually avoided for any length of time, since protein is necessary for healthy life and normal body function. Advice on dietary protein is readily available from any dietician.

In addition, encephalopathy can be improved by drug therapy. It is possible to kill all (or nearly all) the bacteria in the intestine by giving an antibiotic to the patient. Neomycin is the antibiotic normally chosen because it is a powerful drug that is taken by mouth and is not absorbed from the intestine. Another drug which has been used with increasing effect in recent years is lactulose. This works in two different ways. First, it increases the acidity of the bowel contents. The bacteria which inhabit the bowel only thrive under certain conditions and quickly succumb to a change in acidity. Secondly, it also works as a laxative. We know that encephalopathy is made worse by constipation; it is not difficult to understand that, if the bowel contents pass through very slowly, bacteria have a much greater opportunity to break down the protein. The dose of lactulose is increased gradually until diarrhoea develops. While nobody appreciates having diarrhoea all the time, it is usually possible to adjust the dose so that patients have regular, somewhat loose, bowel actions. Lactitol is another drug which works in the same way as lactulose.

In practice, protein restriction and one or more drugs are used to treat patients with encephalopathy. The worst symptoms can usually be controlled but minor features often remain. Close friends and relatives will often note subtle changes in personality in people with cirrhosis, due to encephalopathy.

Oesophageal varices

In Chapter 1, we outlined the anatomy of the blood supply to the liver (p. 3). All the blood from the intestine drains via the portal vein to the liver before passing into the general circulation. Therefore, any disease of the liver or portal vein which obstructs the flow of blood will cause 'back pressure' down the system, resulting in dilated veins (Fig. 16). It is rather like a garden hosepipe with the tap full on at one end and the other end nearly closed—the pressure inside the hose will rise, and any weak points which have appeared over the winter may burst. The body reacts to this problem by opening up many small veins that connect the portal system with the general (systemic) circulation, in order to by-pass the blockage. This is just like the traffic being diverted down the country lanes when there are road works or a crash on the main road. The route is tortuous but eventually the drivers find their way back to the main road beyond the block. However, the country lanes are not designed to carry the heavy traffic and there is congestion and some damage to the hedges and verges! The body's by-passes are in several areas, but the veins that most commonly open up are those around the lower oesophagus and stomach. These by-passes are all performing a useful function, but there is one problem: the veins that form on the inside of the oesophagus can burst and cause severe bleeding. These veins are called 'oesophageal varices' and resemble varicose veins in the legs where high pressure also produces a dilated, tortuous appearance.

When and why do varices bleed?

Some patients have varices for many years and never bleed from them, while others bleed from them very soon after they develop. A bleed may be the first sign of cirrhosis. We do not yet know what precipitates a bleed or how it can be prevented. Varices do not bleed unless the pressure is above a certain level, and in general the bigger the varices the more likely they are to bleed. This is hardly unexpected, but, in

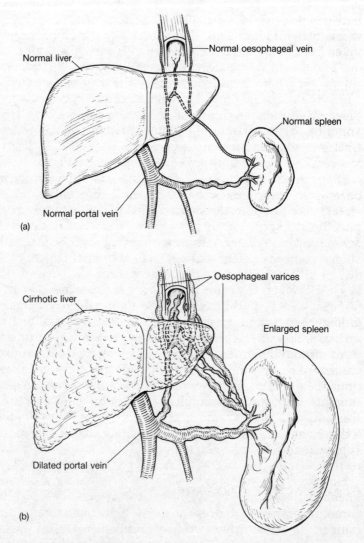

Fig. 16 Some of the changes which take place in portal hypertension. Normally, (a) the spleen is small. When the liver is cirrhotic, (b) and portal hypertension develops, the spleen becomes enlarged and large, tortuous veins appear, especially around the oesophagus.

any individual patient, there is no direct relationship between pressure and bleeding and it is impossible to predict that a patient with a certain pressure will bleed within any given period of time.

What happens when varices bleed?

Bleeding from varices can be very severe and sudden, with vomiting of large quantities of blood and passing black blood (melaena) from the rectum. The patient may collapse with a low blood pressure and rapid pulse and may bleed to death without treatment. Alternatively the bleeding may be a slow ooze—just passing black bowel motions or vomiting black streaks like coffee grounds. Any degree of bleeding in a patient with known portal hypertension must be taken seriously and medical advice sought urgently because a slight ooze may herald a severe, possibly fatal bleed.

How can bleeding varices be controlled?

One third of patients have stopped bleeding by the time they arrive at hospital, one third are bleeding slowly, and one third are still bleeding severely. The site of bleeding is confirmed by passing a gastroscope. This is a flexible fibre-optic instrument which enables the doctor to see the inside lining of the oesophagus and stomach. If a varicose vein in the leg bleeds it can easily be stopped by direct pressure of a thumb or a firm bandage. Unfortunately neither technique is appropriate for a bleeding varicose vein in the lower end of the oesophagus!

However, pressure can be applied by inflating a balloon attached to a tube which is passed down the oesophagus through the mouth. Although not very pleasant for the patient, it is a life-saving temporary measure, while a blood transfusion is given and the patient resuscitated. There are several different tubes named after their inventors, such as 'Sengstaken' and 'Linton'. It can only stay down for a maximum of 24 hours or the pressure can damage the oesophageal wall, so plans for further treatment must be made as soon as possible. If the balloon is merely deflated, bleeding will restart in the majority of patients.

Some drugs such as vasopressin and somatostatin are used to control bleeding temporarily as an alternative to the oesophageal balloon. The drugs are injected into an arm vein and do not have to be put into the portal system.

How can rebleeding be prevented?

The best way to stop a vein bleeding is to tie it or clot it. A third, less direct, method is to reduce the pressure in the varices by creating a large by-pass ('shunting'). All these different methods have been used over the years, but clotting (thrombosing) the veins has become more popular in recent years as the first line of treatment in an emergency. This method was first used over 40 years ago, but has been revived because it is simpler and safer for the patient than a major operation, and because new instruments, such as fibre-optic gastroscopes, have made it technically easier. The principle is the same as injecting varicose veins in the leg. The chemical is very quickly diluted in the bloodstream so that it does not cause clotting elsewhere. This technique controls the bleeding in over 90 per cent of cases for a time, and can be performed under general anaesthetic or with a fibre-optic gastroscope under sedation. For the few patients who continue to bleed, an operation will be required (see below).

What is the long-term treatment?

Once the bleeding is controlled, a decision has to be made about long-term treatment before the patient leaves hospital. This is because a thrombosed vein may reopen later or new veins may form, which, in turn, can bleed. The alternatives are:

1 **Long-term injections** This may involve re-injecting every few weeks until all the veins in the lower oesophagus are clotted. Patients are then checked regularly to identify

new veins so that they can be injected before they bleed. The patient only needs to come up to the hospital for part of a day for these checks, which is not a great hardship every few months. We have patients who have been treated by this method for over ten years.

2 Devascularization operations   There are a number of different forms of this operation devised in parts of the world as far apart as Japan, Egypt, and Britain. Operations are commonly named after the surgeon who invented them. The principle is to tie and divide all veins coming into the upper stomach and lower oesophagus and it requires an incision in the upper abdomen and, sometimes, in the chest as well.

3 Shunt operations   The principle of shunt operations is to make a surgical by-pass between the portal system and the main venous system. To continue the analogy of a traffic jam—if the road through the liver is narrow and congested and the traffic is finding its way through the small country lanes (the varices), the way to take the pressure off the country lanes is to build a brand new dual carriageway by-pass which takes all the traffic easily. These operations have been done for thirty years and take many different forms. They are very successful in preventing rebleeding but have two problems: sometimes the 'shunt' clots off, putting the patient back to square one. Secondly, shunts reduce the liver blood flow, which is probably harmful and may cause deterioration in liver function and lead to encephalopathy (as described earlier).

There is increasing experience to suggest that repeated injections are better than surgery in terms of cost and patient comfort, and for those patients who are not fit for operation injection sclerotherapy is the only option. Nevertheless, in parts of the world where repeated injections cannot be carried out, operation remains a valuable treatment.

Gastric varices

In a small proportion of patients, varicose veins form in the stomach instead of, or as well as, in the oesophagus. These also may bleed but are difficult to control by balloon pressure or injection.

Can bleeding be prevented?

Up to now we have been describing the treatment of veins that have bled, but since this is a life-threatening complication, it is reasonable to consider the prevention of bleeding in the first place. Treatment before a bleed is known as 'prophylactic'.

As we have discussed above (p. 83), it is very difficult to predict which patients will bleed and when. While some patients with varices never have any problem, others, particularly with poor liver function, die of their first bleed. Some years ago a study of prophylactic shunts found that there was no advantage in doing such a major operation if a patient had not bled, because the risk of the operation equalled the risk of bleeding. Injections carry very little risk compared with shunt operations and studies are being carried out to see if prophylactic injections are valuable.

What is the overall outlook with modern treatment once a patient has bled?

If a patient survives the emergency admission, the outlook is quite good, but it depends on the underlying liver disease and the degree of damage to liver function. All studies show that alcoholic patients have a far better prognosis if they stop drinking.

Are there any drugs that can help?

Drugs that lower the blood pressure—known as beta-blockers because they block one of the actions of adrenaline—have been used to treat portal hypertension. They produce a slight reduction in the frequency of bleeding from oesophageal varices, but many people are unable to tolerate these drugs because of side-effects and there is no evidence that they improve survival.

What is the future outlook?

Clearly the emphasis must be to prevent cirrhosis in the first place, because portal hypertension is usually a late complication. Recent research in patients who already have portal hypertension is being directed towards finding factors that will predict whether or not they are likely to bleed. If these can be found, then only those with a high risk of bleeding would require prophylatic treatment. At present injection sclerotherapy is the mainstay of emergency treatment with operations kept as a back-up if required. For long-term management, injection is favoured in Britain and operation in the USA. The place of drugs is still uncertain.

Liver cancer

Primary liver cancer (hepatoma) is a complication of cirrhosis of any kind, and it is recognized that the longer a patient survives with cirrhosis the greater the likelihood of developing hepatoma. Now that we are more able to treat the complications of cirrhosis and patients are surviving longer, so primary cancer is becoming a more common cause of death. Liver cancer will be discussed in greater detail in the next chapter.

10 *Liver cancer*

There is considerable confusion about the term 'liver cancer'. Quite commonly patients say that a relative died of 'cancer of the liver'. Nearly always in Britain, this will mean that cancer had spread to the liver from somewhere else—this was *secondary* cancer in the liver. Primary cancer arising from the liver itself is much rarer in the West but common in Africa and East Asia.

Primary liver cancer

This is one of the commonest primary cancers in the world. It is mostly found in a liver that has cirrhosis for one reason or another and so can be considered a late complication of cirrhosis (Chapter 9). It is common in those parts of the world where hepatitis B is also frequently diagnosed (see map, p. 110) and as we are learning more about hepatitis C it appears that this is also an important cause of primary liver cancer in some areas. However, it is likely that there may be other factors as well as hepatitis viruses. In Africa, a chemical from mould on stored peanuts called 'aflatoxin' is thought to be an important contributory factor. Schistosomiasis (Chapter 13) is also associated with liver cancer. There are a few rare tumours in the liver which may be associated with certain hormonal drugs and cancers can arise from the bile-ducts themselves within the liver. These are termed 'cholangiocarcinomas'.

With this brief introduction, we will now answer a few questions about primary liver cancer.

What symptoms does it give?

In Africa and East Asia it usually causes an increase in the size of the liver with pain due to stretching of the liver capsule, and muscle wasting and ill-health. In the West, it is usually diagnosed when a patient already known to have cirrhosis suddenly deteriorates or it may be an incidental finding at autopsy.

How are the tumours diagnosed?

As with gallstones, the ultrasound scan is an effective and completely painless way of outlining the liver. The CAT scanner can also show the size of tumours very well but it is more expensive. Primary liver cancers commonly produce a protein that can be measured in the blood (alpha-fetoprotein) and this is useful in making the diagnosis.

Can primary tumours be removed surgically?

It is possible in a healthy subject to remove three-quarters of the liver and for the patient to survive. This sometimes has to be done when the liver is badly damaged in a road accident. The remaining liver regenerates rapidly and by 6 weeks is back to its original size. However, we have already noted that primary liver cancer commonly occurs in a cirrhotic liver, and half or one-third of a cirrhotic liver cannot support life and does not regenerate normally. Therefore, surgical removal of a primary liver cancer is only possible when it arises in a normal liver, or in a cirrhotic liver when it is very small.

Can they be treated at all?

Many anti-cancer drugs have been tried on liver cancer, either alone or in combination. Only one or two appear to have any effect, and even with these only a third of patients gain significant benefit. Unfortunately these powerful drugs have unpleasant side-effects such as nausea, vomiting, and loss of hair. The response, if it does occur, is usually quite rapid, so the patients need only be subjected to the unpleasant side-effects of the drugs if there is a good chance of improvement. A few patients may survive for some years by having repeated short courses of treatment. New drugs are being discovered each year, but the main thrust of research must be towards prevention. In this context, hepatitis B vaccines could become the most important cancer-preventing agent in the world.

What is the role of liver transplantation?

Liver transplantation can be used to treat certain tumours, but is only appropriate if there is no spread outside the liver and this is rare, and difficult to confirm before operation. For the role of transplantation in liver disease, see Chapter 12.

Secondary liver cancer

How does secondary cancer spread to the liver?

Cancers anywhere in the body spread by four different routes: (1) local invasion of neighbouring organs; (2) spread to lymphatic 'glands' (this is common in cancer of the breast, for example, which spreads to the 'glands' in the armpits); (3) spread across body cavities, i.e. within the abdominal cavity or within the chest; and (4) spread by the bloodstream to remote parts of the body such as the liver and lungs. As described in Chapter 1, all the blood from the digestive organs passes through the liver before joining the general circulation. The liver is able to filter out cancer cells released into the portal vein, from cancers in the stomach, intestine, or pancreas. In addition, cancer cells can reach the liver via the arteries from anywhere else in the body such as the lungs, which is the commonest site of primary cancer in men in Europe and North America. Not all circulating cancer cells stay in the liver and grow: they are either killed by white cells in the blood or by cells in the liver.

Sometimes the primary cancer is small and does not itself give any symptoms so that secondaries in the liver are the first sign of its presence.

What symptoms do they give?

In the early stages there are no symptoms. Small secondary deposits appear in the liver and grow at varying speeds. They may produce pain by stretching the liver capsule as they enlarge, jaundice by blocking large bile-ducts in the liver, or general ill-health. It is rare to get liver failure from secondary cancer.

How are they diagnosed?

They are diagnosed by blood tests, by ultrasound, and by CAT scan. A liver biopsy can then be taken with a needle to confirm the diagnosis (see Chapter 2). This sometimes gives us a clue to the site of the primary tumour if it is not already known.

How can they be treated?

Once there are multiple secondary tumours in the liver, surgery has little to offer. Occasionally a single secondary is removed with a good result, but usually there are also many small secondaries that may not be obvious at operation. Studies are being done in which an anti-cancer agent is put into the hepatic artery or portal vein, when secondary tumours are very small. It is too early to say whether this treatment has any real benefit—and it certainly has risks. The main problem is that cancers of the digestive organs do not respond well to the available drugs. Secondaries from breast cancer, on the other hand, can be partly controlled by hormone-blocking drugs. In the future, no doubt, more effective drugs will be found.

Radiotherapy is not satisfactory for the liver because the dose and area required to destroy the tumour would also, in most cases, severely damage the liver itself.

The best way to prevent secondaries is to diagnose and remove the primary tumour early—before it has spread. For example, the prognosis of cancer of the colon depends directly on the degree of spread: if the tumour is confined to the mucosa of the colon there is a 75 per cent chance of the patient surviving five years, whereas if it has already spread to the liver at the time of operation, the figure is less than 5 per cent.

11 *Liver disease and children*

Introduction

A number of conditions involving the liver may affect children although, fortunately, serious liver disease is rare in this age group. While some of the common diseases in adults such as virus hepatitis can also affect children, there are two particular groups of disorders which are very much more important during the first few years of life. First, problems may arise because of failure of the liver to develop properly and, secondly, there are a number of conditions which can be inherited. Although these may manifest themselves at any age, they tend to appear during the first few years of life. For reasons which will be described later, it is possible for children to inherit a disease from their parents even though both mother and father may be perfectly healthy with no sign of any disease.

Jaundice at birth

Jaundice is very common in infants during the first few days of life, particularly if they are premature. It usually occurs because the liver has not developed quite enough to deal with the bilirubin produced as a result of breaking down red blood cells (see Chapter 3). It is usually quite mild and lasts for only a few days before resolving on its own as the infant's liver matures. When it is more severe (that is, when the jaundice is more noticeable) special treatment is required. The most widely used form of treatment today is photo-therapy, in which the infant is deliberately exposed to ultraviolet light (or sunlight) for a few hours each day. In the past phenobarbitone was used, because it stimulates enzymes in the liver responsible for clearing bilirubin; but this made the babies rather sleepy. In very severe cases it may be necessary to perform an exchange transfusion in which the baby's blood is removed through blood-vessels in the umbilical cord and exchanged with blood from a healthy blood donor. This type of jaundice does not harm the baby, unless it becomes very severe, in which case there is a risk of brain damage, but with modern treatment it should be completely preventable.

Breast milk jaundice

Occasionally babies who are being breast-fed develop jaundice after about two or three weeks. This is caused by a hormone in the milk which affects the baby's liver. The baby is otherwise well and the jaundice disappears within a few days of stopping breast-feeding. Breast-feeding can then be restarted without any recurrence of the jaundice and the baby does not suffer in any way.

Neonatal hepatitis

Occasionally babies develop jaundice a few weeks after birth and show signs of hepatitis (see Chapter 5). There are many different causes of neonatal hepatitis including a number of rare congenital abnormalities, as well as several viruses such as rubella. In many cases, however, no cause can be found. In most babies the jaundice disappears completely and the child is none the worse for the illness. Nevertheless, such infants are usually referred to hospital for investigation in order to exclude biliary atresia (see below).

Biliary atresia

In this disease the bile-ducts connecting the liver and the duodenum (see Chapter 1) fail to develop properly so bile cannot pass from the liver into the bowel. The reason for this failure is uncertain, but the babies appear to be healthy at birth and jaundice only appears two or more weeks later. It has been suggested that these babies may acquire some virus infection at or shortly after birth, although at present there is no proof for this. Special investigations are needed in order to distinguish biliary atresia from neonatal hepatitis, although unfortunately these usually have to be carried out in hospital. It is important, however, that they are performed early on in a child's life since the treatment of biliary atresia must be started within the first three months of life if it is to be effective.

The treatment involves a very complicated and delicate operation in which the liver surface is cut and tiny ducts are joined to the baby's bowel. This allows the bile to drain from the liver and allows the liver, as well as the rest of the baby, to develop properly. It is a very difficult operation which is not always entirely successful even in the most experienced centres. Delaying the operation beyond the first three months of life will almost certainly result in the liver becoming too damaged for the treatment to be successful. If the operation is delayed or if it is unsuccessful, progressive damage to the liver occurs and this leads to cirrhosis and ultimately to death. The only way in which this can be prevented is by liver transplantation (see Chapter 12) and at present biliary atresia is by far the commonest indication for liver transplantation in children.

Virus hepatitis

Both hepatitis A and hepatitis B are common infections in children and adolescents throughout the world. These conditions are described in greater detail in Chapter 5. In general, virus hepatitis is much milder in children than in adults; most acquire the infection and recover from it without even being aware of having had any significant illness. If newborn babies acquire hepatitis B (see Chapter 5) they are much more likely to develop chronic hepatitis B since their body defence is poorly developed. Hepatitis C does not often affect children but we know that a pregnant woman infected with this virus can transmit it to her unborn child across the placenta.

Gilbert's syndrome

This is a very common condition which is not a true disease because it is really no more than a variation of the norm. A small proportion of otherwise healthy individuals possess a minor abnormality in the liver whereby bilirubin is not properly conjugated and so its clearance from the liver into the bile is impaired (see Chapter 3). As a consequence bilirubin accumulates in the liver and so the person is slightly jaundiced most of the time. The jaundice usually becomes more pronounced if the person is unwell or is not eating properly, and it is under these conditions that it is first recognized. Gilbert's syndrome has to be distinguished from conditions in which the liver is damaged and this can usually be done by fairly simple blood tests. The most important point to remember is that Gilbert's syndrome is entirely harmless. As children grow older the condition tends to disappear.

Rarer conditions

There are a number of less common liver conditions which can affect children. Only a few of these will be mentioned briefly.

Wilson's disease

This is a disorder in which copper accumulates in the body. A normal diet includes a small amount of this metal in a variety of foods such as chocolate and nuts, and whereas copper is normally excreted in the bile, people with Wilson's disease seem to be unable to do this properly. As a result, copper can accumulate and damage the liver. It can also accumulate within the brain and cause brain damage. Wilson's disease is inherited as a recessive genetic trait; this means that in order to have the disease one requires two abnormal genes (Fig. 17). People possessing only one abnormal gene are completely fit and healthy but if two such people marry there is a one in four chance of any child inheriting two abnormal genes (one from each parent), and, consequently, the disease. Boys and girls are equally affected. The frequency of the gene in the normal population is about 1 : 200 so the disease is very rare (200 × 200 × 4 =

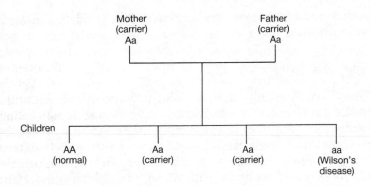

Fig. 17 Illustration of the inheritance of a recessive gene causing Wilson's disease. Each person has two genes, and every child inherits one gene from each parent.

A   Normal gene.

a   Abnormal gene.

AA Normal genetic composition.

Aa Carrier state—these people have one normal and one abnormal gene. They are completely healthy but can pass the abnormal gene on to their children.

aa Wilson's disease—possession of two abnormal genes indicates the disease.

1 : 160 000). Untreated, Wilson's disease causes progressive damage to the liver and brain leading ultimately to death, but effective drug treatment is readily available. Hence, it is important to diagnose and treat the condition as early as possible in life. Therefore, children taken to the doctor with unexplained liver damage are routinely examined and tested for the disease. Since it is inherited, close relatives need to be tested as they may have signs of the condition without being obviously ill. Even though they may be well, they will inevitably go on to develop either liver and/or brain damage without treatment. The most commonly used drug is penicillamine, which acts by binding the excess copper and causing it to be excreted in the urine. The treatment must continue for life, because, if it is stopped, copper will inevitably reaccumulate. With proper treatment, however, children

with Wilson's disease can be restored to good health and lead a normal life.

Alpha-1-antitrypsin deficiency

This is another inherited disorder in which there is faulty production of an enzyme called alpha-1-antitrypsin. The function of this enzyme, which is made in the liver, is to destroy toxic chemicals in the lung and so its absence often leads to lung damage and difficulties with breathing. In addition this condition can cause liver damage, which usually progresses over a number of years leading finally to death from liver failure. Alpha-1-antitrypsin deficiency is inherited in a similar way to Wilson's disease. The gene frequency in the normal population is about 1 in 25, and therefore the disorder affects 1 : 2000. Only a minority of people with the deficiency, however, actually seem to develop liver disease. Liver damage most commonly occurs in infancy or childhood, but sometimes only appears for the first time in later life. Why some individuals develop liver damage and others do not is unknown. At present there is no treatment available. Parents who have had one child with alpha-1-antitrypsin deficiency clearly have a significant chance of having further children with the same condition and they may wish to avoid the risk. It is now possible to diagnose alpha-1-antitrypsin deficiency by taking samples from the fetus during the first few months of pregnancy. This however raises great ethical problems because only some children affected with alpha-1-antitrypsin deficiency will develop any disease.

Reye's syndrome

In this condition children develop an acute illness when large amounts of fat appear in the liver, brain, and kidneys. It may affect fit healthy children at any age from a few months old to mid-teens. It usually produces drowsiness, or even unconsciousness within a day or two, often accompanied by jaundice. Although a small number die, the majority recover completely. The cause is unknown. It sometimes occurs in epidemics, with several cases appearing in a neighbourhood over a short period of time, which suggests that it may be caused by an infection, but so far no single bacterium or virus has yet been implicated. Recently, it has been noticed that many of the children who get Reye's syndrome have been given aspirin shortly before becoming ill and it has been suggested that this may have something to do with the illness. As a result, paediatric aspirins have been withdrawn from sale and parents are advised not to give aspirin to children. Since this has happened, Reye's syndrome has become much less common. The relation between aspirin and this condition is not understood, but it is clearly important that children under the age of 12 must not be given aspirin under any circumstance.

Cystic fibrosis

This an inherited disease which affects mainly the lungs and the pancreas in children, causing recurrent chest infections and diarrhoea. With better treatment most of these people survive into teenage and adult life and a lot of them are developing liver problems. Many develop gallstones, strictures of the bile-ducts, and cirrhosis. There is at present no specific treatment for the liver disease but in a few cases liver transplantation, even combined with lung transplantation, is being carried out.

12 Liver transplantation

The first liver transplant in a patient was performed in 1963. For a long time this was a very difficult operation which was carried out only in a few centres; in the last few years, important advances have greatly improved the success of the procedure so that many more patients are now being treated. In 1990 alone, 1600 liver transplant operations were carried out in the USA, and about 350 were performed in Britain.

Whereas a patient with kidney failure can be maintained on an artificial kidney (dialysis) machine for weeks, months, or even years until the suitable time for transplantation, no such 'artificial liver' is available. This means that the timing of the transplant operation is very critical. Nobody wants to have a major operation with all the risks involved at a time when they are still relatively fit and well. On the other hand, if the surgery is delayed until the patient is very ill with liver failure, then the risk of dying during or shortly after the operation is considerable. In addition, the operation is technically more difficult than kidney transplantation, since the hepatic artery, hepatic and portal veins, and the bile-duct all have to be connected with great precision.

Which patients are suitable for liver transplantation?

In theory, any patient who has advanced liver disease and who is going to die may be considered for a liver transplant. However, it is a very major operation and elderly patients are unlikely to survive the procedure. Similarly, patients with considerable damage to other organs such as the heart or lungs are often unsuitable as the operation will place a further stress on the rest of the body. Babies are not usually accepted for liver transplantation because of the very small blood-vessels which have to be operated on; although successful operations have been carried out on babies less than one year old, much better results are usually obtained in those over the age of two.

The underlying liver condition is also important. Most patients with liver cancer are unsuitable for transplantation since they usually have tumour outside the liver at the time of the transplant. Even if careful investigation shows that the tumour is confined to the liver, most patients develop cancer elsewhere in the body and die within a year or so. In spite of this, occasional long-term survival without cancer recurence is sometimes seen and this justifies carrying out the operation. Whether or not patients with alcoholic cirrhosis should be offered a liver transplant is at present controversial. Some people question whether such an expensive procedure should be offered to people who have self-inflicted disease and argue that successful surgery will merely make the patient well enough to resume his drinking habits. On the other hand, as discussed in Chapter 6, not all patients with alcoholic cirrhosis are necessarily alcohol-dependent and many only realize the damage that alcohol has done to the liver when cirrhosis has reached an advanced stage, and immediately stop drinking.

In summary, there are no absolute contraindications to liver transplantation but patients with non-malignant, non-alcoholic liver disease aged between two and the early sixties, are considered most suitable. Children (those with biliary atresia and certain congenital diseases (Chapter 11)) do appreciably better than adults. Cirrhosis and liver cancer are, however, not the only conditions for which liver transplantation is performed. Some children suffer from rare metabolic disorders in which abnormal enzyme reactions occur in the liver. These can cause abnormalities elsewhere in the body without giving rise to any detectable damage in the liver, but if the liver is replaced with a liver from a normal individual the disorders elsewhere in the body can be cured. An example of this is a very rare type of disturbance in cholesterol production called homozygous hypercholesterol-aemia, where children aged 12–14 can suffer heart attacks due to extremely high blood-cholesterol levels. This can be cured by liver transplant, since the abnormality in cholesterol production is due to an enzyme which acts within the liver.

Acute fulminant liver failure is another indication for liver transplantation. In this, previously fit and healthy people develop a very severe attack of hepatitis due either to a virus (Chapter 5) or to a toxin or poison (Chapter 7) which can be so severe as to be life-threatening. In such patients, death from liver failure will occur unless liver transplantation is carried out urgently. Apart from the medical conditions, however, it must be realized that liver transplantation poses major emotional and domestic demands on both the patient and the family.

How are the livers obtained for transplantation?

Kidney transplants can be carried out using either a live donor (that is, a healthy person gives one of their two kidneys), or a cadaver donor (in which a kidney is removed from someone immediately after death). Naturally, since the entire liver is required, only cadaveric liver transplants are possible. With a few exceptions, any donor whose kidneys are suitable for transplantation will also have a liver which is suitable for transplantation, and both liver and kidney (as well as the heart) can be removed simultaneously and used in different recipients. Although the organs have to be removed within an hour of death, they can be preserved and transported many hundreds of miles to the recipient for transplantation.

Throughout the world, transplant centres have adopted strict rules and regulations to ensure that organs cannot possibly be removed before the donor has suffered brain death.

There is no real shortage of organs available for liver transplantation as far fewer liver transplants are performed than kidney transplants. There are, however, some exceptions to this; patients dying of acute liver failure require urgent transplantation usually within one to two days, and unless there is a very readily available organ they may die before a suitable donor can be found. The other exception concerns children; child donors are essential for the very young since an adult liver is too large to fit in a child's abdomen and although a certain amount of 'trimming down' of a large liver can be performed this is often unsatisfactory.

How can I become an organ donor?

Although donor livers are not at the moment in short supply, there is no cause for complacency since the rapid increase which is taking place in liver transplantation all over the world could soon result in a shortage of available livers. This has already happened in many countries with kidney donors. The continued adequate supply of organs for transplantation depends upon the willingness of healthy people to make provisions to donate their organs for transplantation in the event of sudden death. All you need to do is to obtain a donor card (these are widely available at any hospital or clinic), sign it, and carry it in your wallet or handbag. It is advisable to let your close relatives know your intention to be an organ donor, first so that they do not object (they could be in a position to prevent your organs being used after your death), and second so they can tell the doctors who look after you when you may not be in a position to express your own wishes. In some countries healthy organs may be taken from any patient who dies unless they specifically object, but in most, including Great Britain and Holland, patients have to express their willingness to donate organs. Thanks to willing donors, one sudden, tragic death can result in prolongation of life for three or four other people. Doctors are naturally reluctant to ask permission to remove organs for transplantation from dying children.

What about living donors?

As mentioned above, liver transplantation usually means the donation of the entire cadaver organ. Techniques are being developed, however, whereby the left lobe of the liver together with its bile duct (Fig. 2, p. 3) can be removed from a living donor and transplanted. This would enable adults such as parents to donate part of the liver to children but the technique is very complicated and is still under development.

How long do patients with liver transplants survive?

Survival after liver transplantation is improving every year as more and more experience is being gained. The first three months is the most critical period, during which about one patient in six dies. Another 10 per cent are likely to die during the first year after the operation, but once patients have survived 12 months a large majority are still alive after five years. Those who have survived surgery are usually restored to excellent health, although they require regular follow-up and drug treatment. It is now possible to perform a second transplant operation if the first liver transplant fails for one reason or another.

The survival after emergency liver transplantation for acute liver failure is not so good, since such patients are seriously ill at the time of the operation. Nevertheless, 60 per cent or more of such patients can survive to lead normal, healthy lives.

What can go wrong?

The two main problems are rejection of the transplanted organ, and infection. The body normally rejects tissue that it identifies as not belonging to itself—a part of the immune response. In order for the 'foreign' liver to be accepted, this natural response has to be suppressed by drugs, but unfortunately these drugs also suppress the body's natural defence against infections. Each of these drugs—corticosteroids, azathioprine, and cyclosporin—have other important side-effects.

Can the disease recur in the new liver?

This depends upon the cause of the liver failure. Alcoholic cirrhosis will not recur in the new liver unless the patient resumes drinking. Hepatitis B is not eliminated by transplantation but its recurrence and the damage which it can cause may be substantially reduced by giving the patient hepatitis B globulin (see Chapter 5 p. 52). Similarly, hepatitis C recurs in the transplanted liver, but usually it does not damage the new liver very quickly. There is much debate as to whether primary biliary cirrhosis recurs after liver transplantation; even if it does, there is little or no evidence of serious liver damage during the first five to ten years.

In summary, with the exception of cancer involving the liver, which has been discussed earlier, recurrence of the disease in the new liver occurs so slowly that many years of trouble-free life can usually be anticipated.

What is the demand for, and cost of, liver transplantation?

There is no doubt that liver transplantation is a highly effective operation in carefully selected patients, returning them to a good quality of life for years. The demand for liver transplantation continues to increase as both recognition of suitable patients for the procedure and the technical expertise improves. It cannot be denied, however, that it is an expensive procedure, since not only is the operation itself costly, but the long-term drugs required to prevent rejection are also expensive. This must be compared with the cost of looking after patients with liver failure who need many hospital admissions. Since there are as many as 2000 people a year in the USA who may be suitable for transplantation, this represents an enormous investment in health care. The cost of a liver transplant programme in any country must be borne by either the State or by health insurance organizations, and the most efficient organization is to concentrate the facilities and expertise in a small number of properly equipped centres. There is, however, a limit to the amount of money available for health care and society must decide whether or not this is the best way to spend its budget. It is important to direct research into the causes and prevention of liver disease in its early stages.

13 *Liver disease and the traveller*

International travel is now so common that diseases which were once thought to be confined to exotic corners of the world may now be acquired by holiday-makers on package tours, or by regular travellers in the course of their business trips. A number of these diseases involve the liver. Anyone who develops any illness on returning from a foreign country should make a point of telling the doctor which countries he has visited.

The foreign traveller should take the precaution of enquiring about infectious diseases which occur in the countries he is visiting (even if only for a few days). Protective immunization or treatment is now available against many infections, while in others simple hygienic precautions can be taken to minimize the risk of illness. This advice applies to infections of any sort but here we shall only consider those which involve the liver.

Viral hepatitis

The clinical features of hepatitis are described in detail in Chapter 5. Hepatitis A is endemic throughout the developing countries while hepatitis B is common throughout most of the world. Figure 18 shows the global distribution. 'Endemic' means that the viruses are so common among the population that virtually everybody becomes infected at some time, usually in childhood. The traveller visiting these areas is therefore at great risk of acquiring hepatitis and should take appropriate precautions. The main source of hepatitis A is food and drink. Remember that the local inhabitants in these countries will not be concerned about the risk of hepatitis since they will have acquired it in childhood. Uncooked or improperly cooked food should be avoided if possible and particular care should be taken with drinking water. Although hepatitis viruses can be destroyed by boiling water they will not easily be killed by freezing, so even the ice cubes in your cocktail can be hazardous! Shellfish are recognized as a particularly common source of hepatitis. The risks of acquiring hepatitis are much greater if the living conditions are primitive, but even staying in a five-star hotel does not guarantee protection.

The risk of hepatitis A can be reduced considerably by an injection of gamma globulin. The prospective traveller is advised to have this immediately before he goes abroad and it provides good protection for about 3 or 4 months. People spending longer periods of time in the high-risk parts of the world are advised to have hepatitis A vaccine (p. 51). Although it gives good protection against hepatitis A there is no evidence that it protects against other types of hepatitis.

A much more effective and longer-lasting vaccine against hepatitis B is now available. Details are given in Chapter 5. Since there is no protection for at least 2 months after the first injection the traveller must plan his immunization well in advance of his trip. Not only is hepatitis B vaccination essential for individuals who are likely to be spending a prolonged period of time working in a high-risk part of the world, but it is also now strongly recommended to short-term travellers such as tourists and those on business. It only provides protection against hepatitis B and not against other types of hepatitis viruses.

Other forms of hepatitis (Chapter 5) are also common throughout the world.

Bilharzia (schistosomiasis)

Bilharzia is a very important and serious disease affecting much of South America, Africa, the Middle East, and East Asia (Fig. 19). It is caused by a parasite which lives in freshwater snails. It infects man by penetrating the skin of people who paddle or swim in slow-moving water where the snails live. There are three species of this parasite, but only two of them (*Schistosoma mansoni* and *Schistosoma japonicum*) affect the liver (see Fig. 19). The third type (*Schistosoma haematobium*) affects the bladder. The parasite invades the

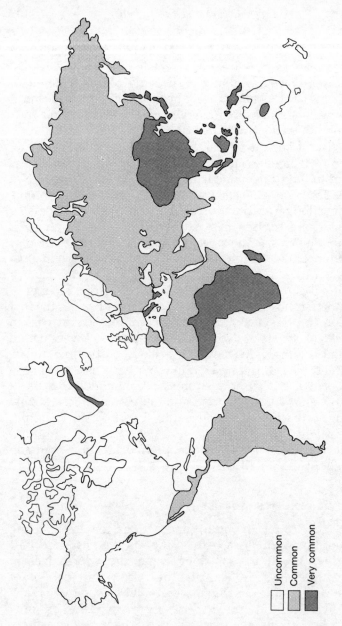

Fig. 18 World-wide distribution of hepatitis B.

Uncommon
Common
Very common

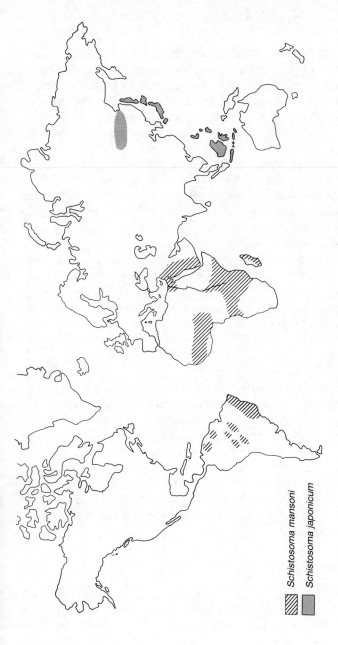

Fig. 19 World-wide distribution of schistosomiasis.

Schistosoma mansoni

Schistosoma japonicum

liver through the bloodstream and then lays its eggs into the branches of the portal vein. This causes damage to the liver, leading to scarring and also to portal hypertension (see Chapter 9). This usually takes several years to develop, so by the time people go to their doctor with their symptoms a considerable amount of damage may already have been done.

Unfortunately bilharzia is a disease which is becoming more, rather than less common in recent years. This is because the widespread development of irrigation schemes, particularly in Africa, has resulted in favourable conditions for the snails which carry the parasite. Many poor countries cannot afford the extra expense to eradicate these snails. The visitor to countries where bilharzia is endemic can avoid the risk simply by not paddling in freshwater areas—or by wearing protective footwear. There is no vaccine against bilharzia but there are a number of drugs available now which can treat the infection.

Amoebiasis

Infection with a parasite called 'Entamoeba histolytica' is a very common cause of dysentery in subtropical and tropical countries throughout the world. In addition it occasionally invades the liver and causes large abscesses. Patients with amoebic liver abscesses are usually very ill with a high fever and a tender enlarged liver; these abscesses may burst and the patient may die unless treated. The parasite is acquired from contaminated food and water. Fortunately amoebic infections can be cured by treatment with the antibiotic, metronidazole.

Hydatid cyst

This is caused by the cyst stage of the dog tapeworm (Echinococcus) which is found all over the world in sheep-rearing communities, but particularly in Australia, South America, Greece, Turkey, and the Middle East. Normally it is transmitted from dogs to sheep, but occasionally man becomes infected by close contact with the sheep dogs (but not the sheep themselves). The eggs invade the liver and form cysts within it. As these cysts become larger in size and in number, they expand the liver and the patient begins to complain of swelling and tenderness in the abdomen. Cysts can also form in other parts of the body including the spleen, lung, and brain. Rupture of a liver cyst into the abdomen causing peritonitis is the most serious complication and this requires an urgent operation. Many people living in sheep-farming areas such as Australia have hydatid cysts for many years without any symptoms. Until recently the only treatment available was an operation in which these cysts were very carefully removed from the liver, but now effective drugs are becoming available.

Liver flukes

These worms are found mainly in East Asia and are acquired from eating raw fish or contaminated watercress. Some of these invade the bile-duct and cause strictures and obstructions which lead to jaundice and fever in a rather similar way to that produced by gallstones (see Chapter 3). Unfortunately they cannot easily be removed by an operation in the same way as gallstones. Drug treatment at present is not very effective for the more serious infections, but in some cases the worms can be killed by drugs.

Malaria

Malarial infection does not cause liver disease itself, although once the parasites infect man they tend to live and multiply in the liver. One of the complications of malaria is haemolysis, and this, if severe, may cause jaundice (see Chapter 4). Many patients with chronic malaria also have enlarged spleens and this may be mistaken for portal hypertension (see Chapter 9). Any traveller visiting an endemic area must take preventative drugs. Because different drugs are needed in different countries, medical advice should be sought.

Leptospirosis

Leptospirosis (Weil's disease) is a very serious infection caused by a spirochaete. This is a spiral-shaped organism which is related to the organism causing syphilis, but leptospirosis is not a venereal disease. Leptospirosis is found throughout the world but the most severe forms of the disease are seen in South-East Asia where it causes an acute illness with deep jaundice, kidney failure, and mental confusion leading to coma. It is transmitted via rats and other vermin and so it is an occupational hazard of people who work in sewers and in rat-infested streams. There is no vaccination against the condition which carries a high mortality when it affects adults, although children seem to be relatively immune.

This catalogue of disease is not meant to spoil the enjoyment of your foreign travel. Most illnesses can be avoided by appropriate vaccination or drugs before travelling, and by sensible precautions about uncooked food and untreated water.

Fig. 20 'The traveller's nightmare'! Most exotic oriental diseases can be avoided.

Appendix: Further information on liver disease

In many countries, charitable organizations have been established to promote interest in liver disease. They help to raise money in order to support research, but in addition they encourage support groups for patients. The names and addresses of five of these organizations are:

United Kingdom   British Liver Trust
182 High Street
Guildford
Surrey
GU1 3HW
Tel. 0483 300869
Fax 0483 66956

United States   American Liver Foundation
1425 Pompton Avenue
Cedar Grove
New Jersey 07009
Tel. (201) 256 2550
Fax (201) 256 3214

Canada   Canadian Liver Foundation
Suite 310
1320 Yonge Street
Toronto
Ontario M4T 1X2
Tel. (416) 964 1953
Fax (416) 964 0024

Germany   Deutsche Leberhilfe e.V.
Gesmolder Strasse 27
D–4520 Melle 1
Germany
Tel. (05422) 6568
Fax (05422) 44499

Children's   Children's Liver Disease Foundation
40–42 Stoke Road
Guildford
Surrey
GU1 4HS
UK
Tel. (0483) 300565
Fax (0483) 300530

Index